Exploring and Settling Our Huge Solar System

Copyright Page

This book is copyrighted for 2021

Exploring and Settling Our Huge Solar System

The Living in Space Series-Book 9

By Martin K. Ettington

ISBN: 9798721144769

Printed in the United States of America

Exploring and Settling Our Huge Solar System

We are all taught in grammar school that the Solar System consists of nine planets (although Pluto's status has changed) and some moons. Maybe the Asteroid Belt is mentioned too.

However, this is a drastic simplification of our Solar System. There are at least 218 natural moons in our system. The Asteroid Belts also include an estimated 1.9 million asteroids. In the outer parts of the system there may be more millions of objects in the Kuiper Belt and even millions more additional objects in the Oort Cloud which is even further out.

In this book we will cover some of the amazing facts about the Moons we know about, many of the amazing objects around our star the Sun, and colonization ideas around the Solar System.

Even though mankind dreams of someday traveling to the nearest stars our Solar System has enough places to visit and colonize for thousands of years.

This book lays out the structures and objects of the Solar System in detail and discusses which parts of it we should colonize, build on, or mine.

Exploring and Settling Our Huge Solar System

Other books by Martin K. Ettington

Spiritual and Metaphysics Books:
Prophecy: A History and How to Guide
God Like Powers and Abilities
Enlightenment for Newbies
Removing Illusions to Find True Happiness
Using the Scientific Method to Study the Paranormal
A Compendium of Metaphysics and How to Guides (Six books together in one volume)
Love from the Heart
The Enlightenment Experience
Learn Your Soul's Purpose
Pursuing Enlightenment
A Modern Man's Search for Truth
Use Intuition and Prophecy to Improve Your Life
The Handbook of Spiritual and Energy Healing

Longevity & Immortality:
Physical Immortality: A History and How to Guide
The Commentaries of Living Immortals
Records of Extremely Long Lived Persons
Enlightenment and Immortality
Longevity Improvements from Science
The 10 Principles of Personal Longevity
Telomeres & Longevity
The Diets and Lifestyles of the Worlds Oldest Peoples
The Longevity Six Books Bundle

Science Fiction:
Out of This Universe
Personal Freedom-Parts 1 & 2
The Psychic Soldier Series:
Book 1-Himalayan Journey
Book 2-A Soldier is Born
Book 3-Fighting For Right
Book 4-Earth Protector
The Immortality Sci Fi Bundle

The God Like Powers Series:
Human Invisibility
Invulnerability and Shielding
Teleportation
Psychokinesis
Our Energy Body, Auras, and Thoughtforms
The God Like Powers Series— Volume 1 Compilation

The Yoga Discovery Series:
Yoga-An Ancient Art Form
Hatha Yoga-Helping you Live Better
Raja Yoga-Through the Ages
The Yoga Discovery Package

Business & Coaching Books:
Creating, Paublishing, & Marketing Practitioner Ebooks
Building a Successful Longevity Coaching Business
Why Become a Coach?
The Professional Coaching Success Trilogy
2020-Make Money Writing and Selling Books
The 2020 Handbook of High Paying Work Without a College Degree

Science, Technology, and Misc.
Future Predictions By and Engineer & Seer
The Unusual Science & Technology Bundle
The Real Atlantis-In the Eye of the Sahara
Real Time Travel Stories From a Psychic Engineer
Removing Limits On Our Consciousness-And Thinking Outside the Box
33 Incredible True Survival Stories
The Importance of Fire in History and Mythology

Exploring and Settling Our Huge Solar System

Exploring and Settling Our Huge Solar System

The Longevity Training Series

(A transcription of the online Multimedia Longevity Coaching Training Program)
The Personal Longevity Training Series-Book1-Long Lived Persons
The Personal Longevity Training Series-Book2-Your Soul's Purpose
The Personal Longevity Training Series-Book3-Enable Your Life Urge
The Personal Longevity Training Series-Book4-Your Spiritual Connection
The Personal Longevity Training Series-Book5-Having Love in Your Heart
The Personal Longevity Training Series-Book6-Energy Body Health
The Personal Longevity Training Series-Book7-The Science of Longevity
The Personal Longevity Training Series-Book8-Physical Body Health
The Personal Longevity Training Series-Book9-Avoiding Accidents
The Personal Longevity Training Series-Book10-Implementing These Principles
The Personal Longevity Training Series-Books One Thru Ten

These books are all available in digital and printed formats from my website and on Amazon, Barnes & Noble, Apple ITunes, and many other sites. My Books Website is: http://mkettingtonbooks.com

Signup for our Mailing List to get the following

1) A discount coupon for 25% discount on all books on our site

2) Occasional Notices of new books available

3) Occasional Email on other offerings of ours (Monthly)

Go to this link to sign-up:

http://personal-longevity.com/mkebooks/emailsignup/

And click this link to get the FREE 102 page Ebook titled "Secrets of Many Things"

If you have any questions about this book or other subjects please contact the Author at:

mke@mkettingtonbooks.com

Exploring and Settling Our Huge Solar System

Table of Contents

1.0 Introduction

The future of Humanity for hundreds of years will be all about settling and mining the Solar System. The objects comprising our Solar System are much vaster than almost anyone realizes.

With high performance propulsion systems like nuclear rockets we will be able to travel to the other planets in weeks and months instead of years.

There are at least 218 natural moons in our system. The Asteroid Belts also include an estimated 1.9 million asteroids. In the outer parts of the system there may be more millions of objects in the Kuiper Belt and even millions more objects in the Oort Cloud which is even further out.

In this book we will cover some of the amazing facts about the Moons we know about and many of the amazing objects around our star, and which ones might make sense to mine or settle in the future.

Even though mankind dreams of someday traveling to the nearest stars our Solar System has enough objects to build on, mine, and colonize for thousands of years.

So be prepared to be amazed about the mysteries of the Solar System and how we can settle it. These opportunities are in our not so distant future. We should start thinking about how we can build civilization and settlements across the Solar System.

2.0   Interesting Facts about our Planets

There is too much to cover to go into technical details about each planet in this book. Therefore, I've mainly presented some key interesting facts about each planet and what we can do with each of them.

2.1   Planet Mercury

Science Fiction often looks at building bases on the terminator of the Sun face and opposite Sun sides of this planet. This is because the same side of Mercury faces the Sun all of the time. While the Sun facing side of Mercury can melt lead, the side in the Sun's shadow is hundreds of degrees below zero, so building a base in between these two places would provide more moderate temperatures.

Mercury does not have any moons or rings.

Mercury is the smallest planet.

Mercury is the closest planet to the Sun.

Your weight on Mercury would be 38% of your weight on Earth.

A solar day on the surface of Mercury lasts 176 Earth days.

A year on Mercury takes 88 Earth days.

It's not known who discovered Mercury.

A year on Mercury is just 88 days long.
One solar day (the time from noon to noon on the planet's surface) on Mercury lasts the equivalent of 176 Earth days while the sidereal day (the time for 1 rotation in relation to a fixed point) lasts 59 Earth days. Mercury is nearly tidally locked to the Sun and over time this has slowed the rotation of the planet to almost match its orbit around the Sun. Mercury also has the

highest orbital eccentricity of all the planets with its distance from the Sun ranging from 46 to 70 million km.

Mercury is the smallest planet in the Solar System.
One of five planets visible with the naked eye, Mercury is just 4,879 Kilometers across its equator, compared with 12,742 Kilometers for the Earth.

Mercury is the second densest planet.
Even though the planet is small, Mercury is very dense. Each cubic centimeter has a density of 5.4 grams, with only the Earth having a higher density. This is largely due to Mercury being composed mainly of heavy metals and rock.

Mercury has wrinkles.
As the iron core of the planet cooled and contracted, the surface of the planet became wrinkled. Scientist have named these wrinkles, Lobate Scarps. These Scarps can be up to a mile high and hundreds of miles long.

Mercury has a molten core.
In recent years scientists from NASA have come to believe the solid iron core of Mercury could in fact be molten. Normally the core of smaller planets cools rapidly, but after extensive research, the results were not in line with those expected from a solid core. Scientists now believe the core to contain a lighter element such as sulphur, which would lower the melting temperature of the core material. It is estimated Mercury's core makes up 42% of its volume, while the Earth's core makes up 17%.

Mercury is only the second hottest planet.
Despite being further from the Sun, Venus experiences higher temperatures. The surface of Mercury which faces the Sun sees temperatures of up to 427°C, whilst on the alternate side this can be as low as -173°C. This is due to the planet having no atmosphere to help regulate the temperature

Mercury is the most cratered planet in the Solar System.
Unlike many other planets which "self-heal" through natural geological processes, the surface of Mercury is covered in craters. These are caused by numerous encounters with asteroids and comets. Most Mercurian craters are named after famous writers and artists. Any crater larger than 250 kilometers in diameter is referred to as a Basin. The Caloris Basin is the largest impact crater on Mercury covering approximately 1,550 km in diameter and was discovered in 1974 by the Mariner 10 probe.

Only two spacecraft have ever visited Mercury.
Owing to its proximity to the Sun, Mercury is a difficult planet to visit. During 1974 and 1975 Mariner 10 flew by Mercury three times, during this time they mapped just under half of the planet's surface. On August 3rd 2004, the Messenger probe was launched from Cape Canaveral Air Force Station, this was the first spacecraft to visit since the mid 1970's.

Mercury is named for the Roman messenger to the gods.
The exact date of Mercury's discovery is unknown as it pre-dates its first historical mention, one of the first mentions being by the Sumerians around in 3,000 BC.

Mercury has an atmosphere (sort of).

Mercury has just 38% the gravity of Earth, this is too little to hold on to what atmosphere it has which is blown away by solar winds. However while gases escape into space they are constantly being replenished at the same time by the same solar winds, radioactive decay and dust caused by micrometeorites.

Exploring and Settling Our Huge Solar System

2.2 Planet Venus

Some scientists and science fiction writers think that maybe Venus could be terraformed to allow us to live there. It would involve finding a way to shield Venus from much of the sunlight it receives to cool it off.

Water could be provided from asteroids around the Solar System. A new place to live in the Solar System?

A day on Venus is longer than a year
It takes Venus longer to rotate once on its axis than to complete one orbit of the Sun. That's 243 Earth days to rotate once - the longest rotation of any planet in the Solar System - and only 224.7 Earth days to complete one orbit of the Sun.

Venus is hotter than Mercury despite being further away from the Sun
Its mean temperature is 462°C. This is because of the high concentration of carbon dioxide in Venus' atmosphere which works to produce an intense greenhouse effect, trapping heat in the atmosphere like a blanket and causing

the temperature of the planet to be much higher than its proximity to the Sun would suggest.

Unlike the other planets in our solar system, Venus spins clockwise on its axis.
All other planets spin anti-clockwise on their axis and orbit the Sun in an anti-clockwise direction. Venus also orbits the Sun anti-clockwise, but its unusual axis rotation is due to being upside down - it was knocked off its upright position earlier in its history! Astronomers believe that at some point, a colliding celestial body tilted Venus so far off its original position that it is now upside down. The only other planet to spin in a weird direction is Uranus which spins on its side, probably the result of another collision early on in its life.

Venus is the second brightest natural object in the night sky after the Moon
The clouds of sulfuric acid in Venus' atmosphere make it reflective and shiny and obscure our view of its surface. Its brightness makes it visible during the day if it's clear and you know where to look

Venus has 90 times the atmospheric pressure of Earth
That's about the same as the pressure at a depth of 1 km in Earth's oceans.

Venus is named after the Roman goddess of love and beauty
It is thought that Venus was named after the beautiful Roman goddess (counterpart to the Greek Aphrodite) due to its bright, shining appearance in the sky. Of the five planets known to ancient astronomers, it would have been the brightest.

Venus was the first planet to have its motions plotted across the sky, as early as the second millennium BC Because Venus is easy to spot with the naked eye, it is impossible to say who discovered the planet. But over the centuries, we have been able to measure Venus' motions, such as the rare transit of Venus when the planet appears from Earth to cross in front of the Sun.

2.3 Planet Earth

The Earth's rotation is gradually slowing
This deceleration is happening almost imperceptibly, at approximately 17 milliseconds per hundred years, although the rate at which it occurs is not perfectly uniform. This has the effect of lengthening our days, but it happens so slowly that it could be as much as 140 million years before the length of a day will have increased to 25 hours.
The Earth was once believed to be the center of the universe.

Due to the apparent movements of the Sun and planets in relation to their viewpoint, ancient scientists insisted that the Earth remained static, whilst other celestial bodies travelled in circular orbits around it. Eventually, the view that the Sun was at the center of the universe was postulated by Copernicus, though this is also not the case.

Earth has a powerful magnetic field.
This phenomenon is caused by the nickel-iron core of the planet, coupled with its rapid rotation. This field protects the Earth from the effects of solar wind.

There is only one natural satellite of the planet Earth.

As a percentage of the size of the body it orbits, the Moon is the largest satellite of any planet in our solar system. In real terms, however, it is only the fifth largest natural satellite.

Earth is the only planet not named after a god.

The other seven planets in our solar system are all named after Roman gods or goddesses. Although only Mercury, Venus, Mars, Jupiter and Saturn were named during ancient times, because they were visible to the naked eye, the Roman method of naming planets was retained after the discovery of Uranus and Neptune.

The Earth is the densest planet in the Solar System.

This varies according to the part of the planet; for example, the metallic core is denser than the crust. The average density of the Earth is approximately 5.52 grams per cubic centimeter.

70% of the Earth's surface is covered in water

When astronauts first went into the space, they looked back at the Earth with human eyes for the first time, and called our home the Blue Planet. And it's no surprise. 70% of our planet is covered with oceans. The remaining 30% is the solid ground, rising above sea level.

Earth is mostly iron, oxygen and silicon

If you could separate the Earth out into piles of material, you'd get 32.1 % iron, 30.1% oxygen, 15.1% silicon, and 13.9% magnesium. Of course, most of this iron is actually down at the core of the Earth. If you could actually get down and sample the core, it would be 88% iron. 47% of the Earth's crust consists of oxygen.

Earth doesn't take 24 hours to rotate on its axis

It's actually 23 hours, 56 minutes and 4 seconds. This is the amount of time it takes for the Earth to completely rotate around its axis; astronomers call this a sidereal day. Now wait a second, that means a day is 4 minutes shorter than we think it is. You'd think that time would add up, day by day, and within a few months, day would be night, and night would be day.

A year on Earth isn't 365 days

It's actually 365.2564 days. It's this extra .2564 days that creates the need for leap years. That's why we tack on an extra day in February every year divisible by 4 – 2004, 2008, etc – unless it's divisible by 100 (1900, 2100, etc)… unless it's divisible by 400 (1600, 2000, etc).

Earth has 1 moon and 2 co-orbital satellites

As you're probably aware, Earth has 1 moon (The Moon). But did you know there are 2 additional asteroids locked into a co-orbital orbits with Earth? They're called 3753 Cruithne and 2002 AA29. We won't go into too much detail about the Moon, I'm sure you've heard all about it.

3753 Cruithne is 5 km across, and sometimes called Earth's second moon. It doesn't actually orbit the Earth, but has a synchronized orbit with our home planet. It has an orbit that makes it look like it's following the Earth in orbit, but it's actually following its own, distinct path around the Sun.

2002 AA29 is only 60 meters across, and makes a horseshoe orbit around the Earth that brings it close to the planet every 95 years. In about 600 years, it will appear to circle Earth in a quasi-satellite orbit. Scientists have suggested that it might make a good target for a space exploration mission.

The Earth is not actually round in shape; in fact it is geoid. This simply means that the rounded shape has a slight bulge towards the equator. So what causes this geoid shape? This happens solely because the rotation of the Earth which causes the bulge around the equator.

The Earth tilts at roughly 66 degrees.

Only 3% water of the earth is fresh, rest 97% salted. Of that 3%, over 2% is frozen in ice sheets and glaciers. Means less than 1% fresh water is found in lakes, rivers and underground.

The Asian Continent covers 30% of the total earth land area, but represent 60% of the world's population.

Each winter there are about 1 septillion (1, 000, 000, 000, 000, 000, 000, 000, 000 or a trillion trillion) snow crystals that drop from the sky.

2.4 Planet Mars

Plans to explore and settle Mars are already moving ahead. Many space probes have visited and landed there. Elon Musk has plans to settle Mars by the mid twenty first century. It would be possible to terraform Mars to make it a shirtsleeve environment but it might take centuries. Adding water from icy planetoids would be one of the first things to do.

See my book "All about Mars Journeys and Settlement" for more details. Here are a couple pictures of what bases on Mars might look like:

The Mars Direct Concept

Space X Concept

Mars and Earth have approximately the same landmass.
Even though Mars has only 15% of the Earth's volume and just over 10% of the Earth's mass, around two thirds of the Earth's surface is covered in water. Martian surface gravity is only 37% of the Earth's (meaning you could leap nearly three times higher on Mars).

Mars is home to the tallest mountain in the solar system.
Olympus Mons, a shield volcano, is 21km high and 600km in diameter. Despite having formed over billions of years, evidence from volcanic lava flows is so recent many scientists believe it could still be active.

Only 18 missions to Mars have been successful.
As of September 2014 there have been 40 missions to Mars, including orbiters, landers and rovers but not counting flybys. The most recent arrivals include the Mars Curiosity mission in 2012, the MAVEN mission, which arrived on September 22, 2014, followed by the Indian

Space Research Organization's MOM Mangalyaan orbiter, which arrived on September 24, 2014. The next missions to arrive will be the European Space Agency's ExoMars mission, comprising an orbiter, lander, and a rover, followed by NASA's InSight robotic lander mission, slated for launch in March 2016 and a planned arrival in September, 2016. Update-the Perseverance Rover landed on Mars February 18, 2021

Mars has the largest dust storms in the solar system.
They can last for months and cover the entire planet. The seasons are extreme because its elliptical (oval-shaped) orbital path around the Sun is more elongated than most other planets in the solar system.

On Mars the Sun appears about half the size as it does on Earth.
At the closest point to the Sun, the Martian southern hemisphere leans towards the Sun, causing a short, intensely hot summer, while the northern hemisphere endures a brief, cold winter: at its farthest point from the Sun, the Martian northern hemisphere leans towards the Sun, causing a long, mild summer, while the southern hemisphere endures a lengthy, cold winter.

Pieces of Mars have fallen to Earth.
Scientists have found tiny traces of Martian atmosphere within meteorites violently ejected from Mars, then orbiting the solar system amongst galactic debris for millions of years, before crash landing on Earth. This allowed scientists to begin studying Mars prior to launching space missions.

Mars takes its name from the Roman god of war.
The ancient Greeks called the planet Ares, after their god of war; the Romans then did likewise, associating the

planet's blood-red colour with Mars, their own god of war. Interestingly, other ancient cultures also focused on colour – to China's astronomers it was 'the fire star', whilst Egyptian priests called on 'Her Desher', or 'the red one'. The red colour Mars is known for is due to the rock and dust covering its surface being rich in iron.

There are signs of liquid water on Mars.
For years Mars has been known to have water in the form of ice. The first signs of trickling water are dark stripes or stains on crater wall and cliffs seen in satellite images. Due to Mars' atmosphere this water would have to be salty to prevent it from freezing or vaporizing.

One day Mars will have a ring.
In the next 20-40 million years Mars' largest moon Phobos will be torn apart by gravitational forces leading to the creation of a ring that could last up to 100 million years.

Sunsets on Mars are blue.
During the Martian day the sky is pinkish-red, this is the opposite of the Earth's skies.

Exploring and Settling Our Huge Solar System

2.5   Planet Jupiter

Jupiter's gravity is too high to settle on (2.5 times that of Earth) and it has a huge atmosphere which would crush any spacecraft trying to land on it. However, there have been many proposals to build cities or outposts floating in the Jovian atmosphere. Here is a picture of one concept below:

Picture of a floating city in Jovian Atmosphere

Jupiter is the fourth brightest object in the solar system.
Only the Sun, Moon and Venus are brighter. It is one of five planets visible to the naked eye from Earth.

The ancient Babylonians were the first to record their sightings of Jupiter.
This was around the 7th or 8th century BC. Jupiter is named after the king of the Roman gods. To the Greeks, it represented Zeus, the god of thunder. The Mesopotamians saw Jupiter as the god Marduk and patron of the city of Babylon. Germanic tribes saw this planet as Donar, or Thor
.

Jupiter has the shortest day of all the planets.
It turns on its axis once every 9 hours and 55 minutes. The rapid rotation flattens the planet slightly, giving it an oblate shape.

Jupiter orbits the Sun once every 11.8 Earth years.
From our point of view on Earth, it appears to move slowly in the sky, taking months to move from one constellation to another.

Jupiter has unique cloud features.
The upper atmosphere of Jupiter is divided into cloud belts and zones. They are made primarily of ammonia crystals, sulfur, and mixtures of the two compounds.

The Great Red Spot is a huge storm on Jupiter.
It has raged for at least 350 years. It is so large that three Earths could fit inside it.

Jupiter's interior is made of rock, metal, and hydrogen compounds.
Below Jupiter's massive atmosphere (which is made primarily of hydrogen), there are layers of compressed hydrogen gas, liquid metallic hydrogen, and a core of ice, rock, and metals.

Jupiter's moon Ganymede is the largest moon in the solar system.
Jupiter's moons are sometimes called the Jovian satellites, the largest of these are Ganymede, Callisto, Io, and Europa. Ganymede measures 5,268 km across, making it larger than the planet Mercury.

Jupiter has a thin ring system.
Its rings are composed mainly of dust particles ejected from some of Jupiter's smaller worlds during impacts from incoming comets and asteroids. The ring system begins some 92,000 kilometers above Jupiter's cloud tops and stretches out to more than 225,000 km from the planet. They are between 2,000 to 12,500 kilometers thick.

Eight spacecraft have visited Jupiter.
Pioneer 10 and 11, Voyager 1 and 2, Galileo, Cassini, Ulysses, and New Horizons missions. The Juno mission arrived at Jupiter July 2016. Other future missions may focus on the Jovian moons Europa, Ganymede, and Callisto, and their subsurface oceans.

2.6   Planet Saturn

Saturn is also a gas giant so it could not be settled on land but floating bases or cities like those in the Jupiter section are possible.

There might also be a chance to mine the rings of Saturn for water ice.

Saturn is huge. It is the second largest planet in our Solar System. Jupiter is the only planet that is bigger.

You cannot stand on Saturn. It is not like Earth. Saturn is made mostly of gases. It has a lot of helium. This is the same kind of gas that you put in balloons.

Its beautiful rings are not solid. They are made up of bits of ice, dust and rock.

Saturn is the only planet that could float in water. That would take a really big bath tub!

Some of these bits are as small as grains of sand. Some are much larger than tall buildings. Some are up to a kilometer (more than half-a-mile) across.

The rings are huge but thin. The main rings could almost go from Earth to the moon. Yet, they are less than a kilometer thick.

Other planets have rings. Saturn's rings are the only ones that can be seen from Earth. All you need is a small telescope.

Saturn could float in water because it is mostly made of gas. (Earth is made of rocks and stuff.)

It is very windy on Saturn. Winds around the equator can be 1,800 kilometers per hour. That's 1,118 miles per hour! On Earth, the fastest winds "only" get to about 400 kilometers per hour. That's only about 250 miles per hour.

Saturn goes around the Sun very slowly. A year on Saturn is more than 29 Earth years.

Saturn spins on its axis very fast. A day on Saturn is 10 hours and 14 minutes.

The Ringed Planet is so far away from the Sun that it receives much less sunlight than we do here on Earth. Yes, the Sun looks smaller from there.

The day Saturday was named after Saturn.

2.7   Planet Neptune

It takes Neptune 164.8 Earth years to orbit the Sun. On 11 July 2011, Neptune completed its first full orbit since its discovery in 1846.

Neptune was discovered by Jean Joseph Le Verrier. The planet was not known to ancient civilizations because it is not visible to the naked eye. The planet was initially called Le Verrier after its discoverer. This name, however, quickly was abandoned and the name Neptune was chosen instead.

Neptune is the Roman God of the Sea. In Greek, Neptune is called Poseidon.

Neptune has the second largest gravity of any planet in the solar system – second only to Jupiter.

The orbit path of Neptune is approximately 30 astronomical units (AU) from the Sun. This means it is around 30 times the distance from the Earth to the Sun.

The largest Neptunian moon, Triton, was discovered just 17 days after Neptune itself was discovered. Neptune has a storm similar the Great Red Spot on Jupiter. It is commonly known as the Great Dark Spot and is roughly the size of Earth.

Neptune also has a second storm called the Small Dark Spot. This storm is around the same size as Earth's moon.

Neptune spins very quickly on its axis. The planets equatorial clouds take 18 hours to complete one rotation. The reason this happens is that Neptune does not have a solid body.

Only one spacecraft, the Voyager 2, has flown past Neptune. It happened in 1989 and captured the first close-up images of the Neptunian system. It took 246 minutes – four hours and six minutes – for signals from Voyager 2 to reach back to Earth.

The climate on Neptune is extremely active. In its upper atmosphere, large storms sweep across it and high-speed solar winds track around the planet at up to 1,340 km per second. The largest storm was the Great Dark Spot in 1989 which lasted for around five years.

Like the other outer planets, Neptune possesses a ring system, though its rings are very faint. They are most likely made up of ice particles and grains of dust with a carbon-based substance coating them.

Neptune has 14 known moons. The largest of these moons is Titan – a frozen world which spits out particles of nitrogen ice and dust from below its surface. It is believed that Titan was caught by the immense gravitational pull of

Neptune and is regarded as one of the coldest worlds in our solar system.

Neptune has an average surface temperature of -214°C – approximately -353°F.

2.8 Planet Uranus

Uranus was officially discovered by Sir William Herschel in 1781.
It is too dim to have been seen by the ancients. At first Herschel thought it was a comet, but several years later it was confirmed as a planet. Herscal tried to have his discovery named "Georgian Sidus" after King George III. The name Uranus was suggested by astronomer Johann Bode. The name comes from the ancient Greek deity Ouranos.

Uranus turns on its axis once every 17 hours, 14 minutes.
The planet rotates in a retrograde direction, opposite to the way Earth and most other planets turn.

Uranus makes one trip around the Sun every 84 Earth years.
During some parts of its orbit one or the other of its poles point directly at the Sun and get about 42 years of direct sunlight. The rest of the time they are in darkness.

Uranus is often referred to as an "ice giant" planet.
Like the other gas giants, it has a hydrogen upper layer, which has helium mixed in. Below that is an icy "mantle, which surrounds a rock and ice core. The upper atmosphere is made of water, ammonia and the methane ice crystals that give the planet its pale blue colour.

Uranus hits the coldest temperatures of any planet.
With minimum atmospheric temperature of -224°C Uranus is nearly the coldest planet in the solar system. While Neptune doesn't get as cold as Uranus it is on average colder. The upper atmosphere of Uranus is covered by a methane haze which hides the storms that take place in the cloud decks.

Uranus has two sets of very thin dark colored rings.
The ring particles are small, ranging from a dust-sized particles to small boulders. There are eleven inner rings and two outer rings. They probably formed when one or more of Uranus's moons were broken up in an impact. The first rings were discovered in 1977 with the two outer rings being discovered in the Hubble Space Telescope images between 2003 and 2005.

Uranus' moons are named after characters created by William Shakespeare and Alexander Pope.
These include Oberon, Titania and Miranda. All are frozen worlds with dark surfaces. Some are ice and rock mixtures. The most interesting Uranian moon is Miranda; it has ice canyons, terraces, and other strange-looking surface areas.

Only one spacecraft has flown by Uranus.
In 1986, the Voyager 2 spacecraft swept past the planet at a distance of 81,500 km. It returned the first close-up images of the planet, its moons, and rings.

2.9 Planet Pluto

A deep space base on Pluto might make sense since it has one fifteenth of Earth's gravity. It would be easy to land and take off and have a large surface to build bases.

A Painting of what the ground on Pluto might be like

Pluto is named after the Roman god of the underworld. This was proposed by Venetia Burney an eleven year old schoolgirl from Oxford, England.

Pluto was reclassified from a planet to a dwarf planet in 2006.
This is when the IAU formalized the definition of a planet as "A planet is a celestial body that (a) is in orbit around the Sun, (b) has sufficient mass for its self-gravity to overcome rigid body forces so that it assumes a hydrostatic equilibrium (nearly round) shape, and (c) has cleared the neighborhood around its orbit."

Pluto was discovered on February 18th, 1930 by the Lowell Observatory.
For the 76 years between Pluto being discovered and the time it was reclassified as a dwarf planet it completed under a third of its orbit around the Sun.

Pluto has five known moons.
The **moons** are Charon (discovered in 1978,), Hydra and Nix (both discovered in 2005), Kerberos originally P4 (discovered 2011) and Styx originally P5 (discovered 2012) official designations S/2011 (134340) 1 and S/2012 (134340)

Pluto is the largest dwarf planet.
At one point it was thought this could be Eris. Currently the most accurate measurements give Eris an average diameter of 2,326km with a margin of error of 12km, while Pluto's diameter is 2,372km with a 2km margin of error.

Pluto is one third water.
This is in the form of water ice which is more than 3 times as much water as in all the Earth's oceans, the remaining two thirds are rock. Pluto's surface is covered with ices, and has several mountain ranges, light and dark regions, and a scattering of craters.

Pluto is smaller than a number of moons.
These are Ganymede, Titan, Callisto, Io, Europa, Triton, and the Earth's moon. Pluto has 66% of the diameter of the Earth's moon and 18% of its mass. While it is now confirmed that Pluto is the largest dwarf planet for around 10 years it was thought that this was Eris.

Pluto has an eccentric and inclined orbit.
This takes it between 4.4 and 7.3 billion km from the Sun meaning Pluto is periodically closer to the Sun than Neptune.

Pluto has been visited by one spacecraft.
The *New Horizons* spacecraft, which was launched in 2006, flew by Pluto on the 14th of July 2015 and took a series of images and other measurements. *New Horizons* is now on its way to the Kuiper Belt to explore even more distant objects.

Pluto's location was predicted by Percival Lowell in 1915.
The prediction came from deviations he initially observed in 1905 in the orbits of Uranus and Neptune.

Pluto sometimes has an atmosphere.
When Pluto's elliptical orbit takes it closer to the Sun, its surface ice thaws and forms a thin atmosphere primarily of nitrogen which slowly escapes the planet. It also has a methane haze that hovers about 161 kilometers above the surface. The methane is dissociated by sunlight into hydrocarbons that fall to the surface and coat the ice with a dark covering. When Pluto travels away from the Sun the atmosphere then freezes back to its solid state.

Exploring and Settling Our Huge Solar System

3.0 The Moons of the Planets

In this chapter we will review the most interesting moons of the planets. For a current list of all the moons in the Solar System see the Appendix at the end of the book.

Aside from the Moon orbiting the Earth the first moon found around other planets were the Galilean moons first observed around Jupiter in January 1610 by Galileo Galilei. He observed that they were orbiting the giant planet of Jupiter. These four moons are Io, Europa, Ganymede, and Callisto.

The next discovery was the moon of Titan circling Saturn in 1655 by Christiaan Huygens who was a Dutch physicist, mathematician, astronomer and inventor.

Other moons followed over the centuries with the pace picking up significantly in the 20th and early 21st centuries.

Some objects are also classified as dwarf planets instead of just moons or asteroids.

3.1 Earth's Moon

The Earth's Moon is of course the one we know best. It has been in our sky since before man existed. Now we are looking to go back to the Moon and build a moon base at the South Pole. Why? Because lunar satellites have found water ice at the poles. We may be able to mine this ice for rocket fuel, oxygen, and water.

See my book "All about Moon Bases" to learn more about our plans to build bases on the Moon and Space Stations in orbit.

Some interesting facts about our Moon:

The Moon is Earth's only permanent natural satellite
It is the fifth-largest natural satellite in the Solar System, and the largest among planetary satellites relative to the size of the planet that it orbits.

The Moon is the second-densest satellite
Among those whose densities are known anyway. The first densest is Jupiter's satellite Io.

The Moon always shows Earth the same face
The Moon is in synchronous rotation with Earth. Its near side is marked by large dark plains (volcanic 'maria') that fill the spaces between the bright ancient crustal highlands and the prominent impact craters.

The Moon's surface is actually dark
Although compared to the night sky it appears very bright, with a reflectance just slightly higher than that of worn asphalt. Its gravitational influence produces the ocean tides, body tides, and the slight lengthening of the day.

The Sun and the Moon are not the same size
From Earth, both the Sun and the Moon look about same size. This is because, the Moon is 400 times smaller than the Sun, but also 400 times closer to Earth.

The Moon is drifting away from the Earth
The Moon is moving approximately 3.8 cm away from our planet every year.

The Moon was made when a rock smashed into Earth
The most widely-accepted explanation is that the Moon was created when a rock the size of Mars slammed into Earth, shortly after the solar system began forming about 4.5 billion years ago.

The Moon makes the Earth move as well as the tides
Everyone knows that the Moon is partly responsible for causing the tides of our oceans and seas on Earth, with the Sun also having an effect. However, as the Moon orbits the Earth it also causes a tide of rock to rise and fall in the same way as it does with the water. The effect is not as dramatic as with the oceans but nevertheless, it is a measurable effect, with the solid surface of the Earth moving by several centimeters with each tide.

The Moon has quakes too
They're not called earthquakes but moonquakes. They are caused by the gravitational influence of the Earth. Unlike quakes on Earth that last only a few minutes at most, moonquakes can last up to half an hour. They are much weaker than earthquakes though.

There is water on the Moon!
This is in the form of ice trapped within dust and minerals on and under the surface. It has been detected on areas of the lunar surface that are in permanent shadow and are

therefore very cold, enabling the ice to survive. The water on the Moon was likely delivered to the surface by comets. The areas of the Moon with water ice are the North and South Poles.

3.11 The Story of Building a Moon Base

My book "The Moon and Beyond" is still a work in progress but it includes a few chapters with the most realistic description I could provide of what the process of building a Moon Base would be like. The following chapters describe an early Moon Base covered by lunar regolith:

<u>Early Moon Base Construction</u>

We landed after several orbital corrections and a powered descent. The landing was anti-climactic with our engine kicking up lots of dust and then the motion stopped and we became aware that we were experiencing a one sixth Earth gravity. After making sure all the systems were operating properly, we opened the main hatch with us all wearing spacesuits to get outside.

The Commander and his assistant went down the ladder. Their first action was to activate the inflatable temporary structure at the base of the lander. This structure started to inflate and would provide us with shelter for the next several weeks as the main base was built. The inflatable dome was twenty feet in diameter and had an airlock built in. We each had a little sleeping cubicle and there was a galley and work areas. After it inflated we went inside to check airflow, heat, and then moved in supplies for living there.

We needed this shelter since living inside the lander for an extended period of time was a guarantee for crew stress and awful overcrowding.

We all took a walk around the landing site since we were all so excited to be there. The landing site was about one half mile from the crater wall and was in deep shade which was a couple of hundred degrees below zero Celsius. I could see the crater rim curving away to the horizon where it went out of sight. On the other side of the lander the land was just flat although we could see impact rocks in the distance. We could also see a glow over the crater rim because the Sun was shining on it from the other side. I was also thinking about the best location for the shelter. Did we want it out in the open or next to the crater's wall? Next to the crater wall would be more protected in the long run. However, I didn't need to think long because the building's site had already been picked out on Earth.

That night we had a little party in the temporary structure and all rested well before work was to start in the morning. Next morning Olga and I were the first ones outside. We needed to prepare the shelter site and setup the 3D construction equipment. There was a small tractor to be assembled which would be used to dig out a base for the building and make sure the foundation was firm. The tractor used a radio isotopic power source using Plutonium to create heat which was converted to electricity. The tractor had a radio control which we had practiced with back on Earth to control its movements.

I spent the next few hours digging a foundation pit with Olga relieving me as needed. After several days of effort we had a sufficient foundation dug and ready for construction. The foundation was round and one hundred feet in diameter. Our intent was to build a structure which could eventually hold fifty living and working people inside. The next step was the construction of the larger girders for the three dimensional construction machine. Before we could actually start building the building we needed girders to raise the machine above the ground and provide tracks to move the construction printing head over the construction. Imagine that we were building a large printer larger than the building size. The printing head would move in computer controlled movements over the ground to print the building underneath it. The preparation project took several days.

We had been on the Moon a week and we were dead tired at the end of each workday. The commander made sure we all ate dinner together in the temporary shelter to update each other and build a sense of community.

Captains Hold and Neemar had just gotten back from a field trip up to the crater rim. Their job was to install a communications antenna with repeaters aimed to our site back inside the crater. They used an open rocket powered vehicle to launch up to the rim and come back. At the top in the sunlight, and in line with Earth they installed solar panels and the antenna and communications equipment we needed to have regular communications with the DSG and Earth.

Stark and Springer were responsible for the ice mining. They had already done radar surveys within a couple miles radius of our landing site using a simple lunar rover and were now laying out the foundation for the mining and extraction site. Another mission would bring sufficient mining and purification equipment to start the generation of large quantities of water, and its extraction into hydrogen and oxygen.

<u>Main Shelter Construction</u>

Finally, after almost two weeks of foundation digging, pre-construction, and setup of the 3D printing equipment we were ready to get started.

The machine had a hopper where we would feed it with Moon materials of a granular type, and a silicate based binder which would bind it all together like concrete. We had the machine programmed to build a three level building with living quarters on the lowest (and safest) level, with office, labs, and manufacturing on the upper levels. There was even a garden area to grow vegetables to enhance our diet and produce some oxygen and filter out some carbon dioxide.

We turned on the machine and it started printing the bottom level of the building. All we had to do was keep feeding it the raw lunar regolith materials, the silicate binder, and electrical and piping which it would place and hold in locations as it poured walls and those materials became locked in place by the hardening walls. We wanted more solid materials like rebar to support the structure but didn't have the capacity to carry those materials to the Moon. Instead we were counting on the walls to be hard enough and carry enough load to make the structure solid.

Our building would be like those built by the Romans—who invented concrete. The Romans would use volcanic ash and lime in their concrete which they used to build many seaports and famous building like the Roman Colosseum. The Romans also didn't use rebar and most of their concrete structure were pure concrete like the Pantheon which was built with different thicknesses of concrete and well thought out geometric designs to give it the strength to hold up for two thousand years.

Over the next several days the building machine first printed a floor for the whole structure, then we could see the walls rising on the basement level which was designed to be twelve feet tall to give a sense of space. It would also have a hanging ceiling with air and other utilities in it which would make the visible height ten feet.

The plan was to finish and roof over the basement with its own airlock entrance as the rest of the building was completed. An elevator shaft was installed but blocked off temporarily. A ramp actually led up to the airlock on the first floor.

As the first level was completed Olga and I started moving in environmental systems and connecting them up. First was the air generation and ventilating system. It worked off of water ice which was now being produced from the ice mine. An automated supply ship also landed on Day 25. It homed in on a beacon our people had planted several hundred feet from our main base. The supply ship contained the rest of the initial ice mining and electrolysis equipment to produce usable quantities of fuel and other components we needed like air.

We also hooked up the waste recycling system. This system would dry out human waste and recycle the liquids. The dried waste could be used as fertilizer in the garden. It would be nice to use a toilet again rather than the tubes and bodily waste connections in the temporary inflatable shelter. We had to cut down the percentage of oxygen in the air to reduce fire risks and had brought tanks of liquid nitrogen which we installed with the air equipment to reduce the oxygen percentage to only twenty percent.

Each day Olga and I would take turns so that one of us was monitoring the construction machine while the other was working on systems in the building basement. After two more weeks the basement had a heater installed and air working inside. It also had a six inch thick ceiling to keep out solar radiation.

This was fortunate because the Commander called us all together early before dinner and we wondered what was up. He told us Earth was advising us of a large solar storm which would hit within six hours. It was projected to last several days and the inflatable tent would not provide enough protection. We would basically be barbequed if we stayed in the tent.

The main option was for us to live in the lander which was so compact nobody really wanted to do it. Both Olga and I suggested the basement of our structure was ready for occupation and would be safer than the lander because of its six inch thick ceiling. It would be ideal for a larger solar flare to have our full three foot thick roof. But given the projections of the flare, the current roof should work. Also due to the Moon's angle to the Sun and orbit around the Earth, we wouldn't be exposed to radiation problems for most of the storm. The Commander asked us more questions to be sure of the safety but we could see the relief on his face that we would not need to go live in the lander again for days.

Pretty soon he was assigning everyone tasks to move our food, sleeping equipment and more to the main shelter's basement. Over the next few hours we were a beehive of activity as we moved everything we could into the basement of the in progress building. That night we were all in the unfinished basement and it was pretty messy, but all of our life support systems were working properly. We had radiation monitors all over the basement and the only area which registered dangerous was out next to the airlock upstairs.

We spent the next several days playing cards and watching movies while waiting for the solar storm to finish. The next morning we were given the all clear by Earth and resumed construction. Now you could see the outline of the walls on the main level as the printed building continued to grow.

Finishing the Shelter

Olga and I restarted construction on the building that day. She worked outside while I worked on the interior of the basement level. This included setting up partitions for rooms. The pre-fab partitions were constructed outside by the construction machine. Then I would take them inside and position them for the rooms. They would then snap together to form walls and even doorways. The walls would fit in slots in the floor which were part of the original construction. I only had to do some drilling and screwing to connected power, doors, and more for each room.

As the rooms were built everyone started moving their personal items into them. Then I also started working on the kitchen and galley area. I got some help from other crew members who wanted this finished as quickly as possible.

Going outside that afternoon I could see that the walls were rising well on the main floor which contained some pre-defined rooms. The rest of the rooms would be based on movable partitions.

Away from the main shelter construction continued. The ice mining operation and splitting into component oxygen and hydrogen was now looking pretty close to completion. Over the next few weeks the main floor of the shelter was finished and the smaller third floor was now under construction. You could see the main roof taking shape as the third floor grew. After another week of construction the third floor overall structure and roof was completed. The structure had no windows to keep the interior radiation safe. Windows would be simulated from exterior cameras which could display on large window screens inside.

My next big task was to run the tractor to push regolith over the roof. The idea was to bury it with at least several feet of covering to provide full protection well beyond any type of projected solar flare we could imagine.
I started by building a ramp on the side of the shelter away from the two airlocks. It took me most of that day to build the ramp. The next week was all about plowing regolith onto the roof. The regolith was then compacted in place by the building machine.

Finally, after over a month of construction you could look at the full outline of main shelter. It didn't look like much from the outside. All you could see from there was a big pile of soil with an entrance ramp and ramp cover going into a dark cavity. There were actually airlock ramps on two sides of the structure. In case one was needed as an emergency entrance.

When you entered the airlock you waited for the air pressure to be equalized. Then the door would open and you would enter into the equipment room with racks for spacesuits, and lockers with other outside equipment. Then a pressure tight door opened into the main shelter. At this location there were stairs up to the second level or down into the basement. A freight elevator was also next to the stairs to take larger equipment up and down. While the basement living area was pretty much finished, there was still a lot of construction on levels one and two. This construction would go on for months and the next crew rotation would also be continuing building.

3.2   The Moons of Mars

Mars has two natural moons: Phobos and Deimos

 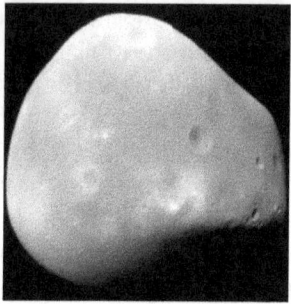

Pictures of Phobos and Deimos

Phobos and Deimos might be great places to build orbiting stations around Mars with their low gravity and as ship docking platforms already in orbit around the planet.

A Ship Docking Platform and Base on one of the Moons

They are irregular in shape. Both were discovered by American astronomer Asaph Hall in August 1877 and are named after the Greek mythological twin characters Phobos (fear) and Deimos (panic) who accompanied their father Ares into battle. Ares, god of war, was known to the Romans as Mars.

Johnathon Swift wrote about the moons of Mars in Gulliver's Travels, published in 1726, and in the book he said that there were two moons and exactly described their size and orbital period. The implication is that Swift somehow knew about the moons of Mars, Phobos and Deimos, a century and a half before they were discovered. It turns out that his predictions, if you can call them that, were accurate enough to be an interesting coincidence, but not so close that we have to consider it anything but that.

Compared to the Earth's Moon, the moons Phobos and Deimos are small. Phobos has a diameter of 22.2 km (13.8 mi) and a mass of 1.08×10^{16} kg, while these measures for Deimos are 12.6 km (7.8 mi) and 2.0×10^{15} kg. Phobos orbits closer to Mars, with a semi-major axis of 9,377 km (5,827 mi) and an orbital period of 7.66 hours; the semi-major axis of Deimos's orbit is 23,460 km (14,580 mi), with an orbital period of 30.35 hours.

3.3 The Moons of Jupiter

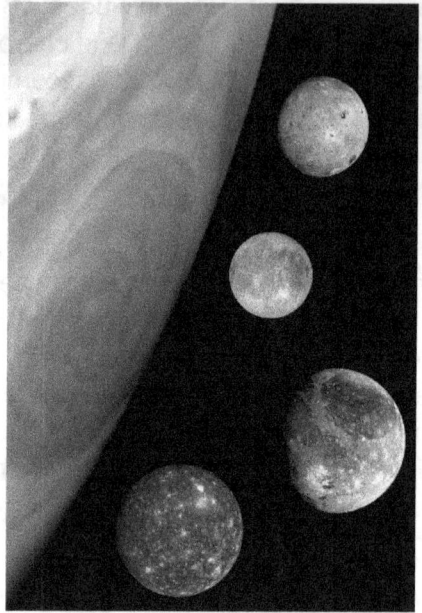

There are over 79 natural moons in the Jupiter system.

The Galilean moons (or Galilean satellites are the four largest moons of Jupiter—Io, Europa, Ganymede, and Callisto. They were first seen by Galileo Galilei in December 1609 or January 1610, and recognized by him as satellites of Jupiter in March 1610. They were the first objects found to orbit a planet other than the Earth.

They are among the largest objects in the Solar System with the exception of the Sun and the eight planets, with radii larger than any of the dwarf planets. Ganymede is the largest moon in the Solar System, and is even bigger than the planet Mercury, though only around half as massive. The three inner moons—Io, Europa, and Ganymede—are in a 4:2:1 orbital resonance with each other. While the

Galilean moons are spherical, all of Jupiter's much smaller remaining moons have irregular forms because of their weaker self-gravitation.

The Galilean moons were observed in either 1609 or 1610 when Galileo made improvements to his telescope, which enabled him to observe celestial bodies more distinctly than ever.

Galileo's observations showed the importance of the telescope as a tool for astronomers by proving that there were objects in space that cannot be seen by the naked eye. The discovery of celestial bodies orbiting something other than Earth dealt a serious blow to the then-accepted Ptolemaic world system, a geocentric theory in which everything orbits around Earth.

A description of these four moons follows:

IO

JUPITER'S MOON IO

The Voyager spacecraft took the first close-up images of Io more than 300 years after the moon's discovery. The images showed a surface with no signs of craters from past impacts. What we saw instead was a surface almost entirely covered with large volcanoes. Cameras on Voyager actually captured volcanic eruptions in progress. The frequency of these sulfuric eruptions has filled in almost all of the impact craters and left Io with one of the youngest looking surfaces in the solar system.

Close-up photos of eruptions in progress show powerfully hot lava glowing orange and red. Photos taken on the night side of Io show not only the hot volcanic vents, but also a thin sulfur dioxide atmosphere produced by constant outgassing. Io's unusual red and orange colors come primarily from sulfur, which condenses on the surface after being outgassed by the volcanoes.

Although there is no direct evidence of tectonic activity on Io, scientists feel confident it exists since the processes that fuel volcanism also fuel tectonics. The volcanic eruptions are so frequent and cover the surface so thoroughly that any clear evidence of tectonic activity is likely to be buried.

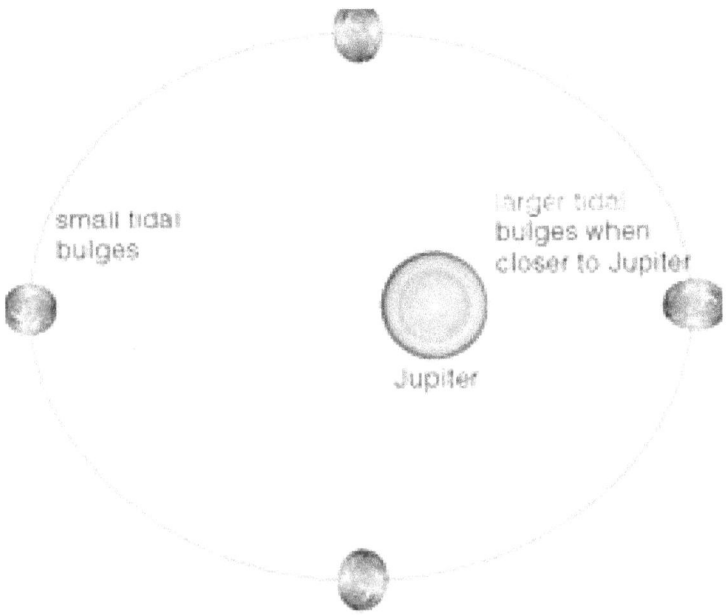

Io's Tidal Bulging
*The bulges and orbital eccentricity
are exaggerated in this diagram.*

Tidal Heating

Io's activity is generated by heat deep inside its center. The force needed to keep Io in synchronous rotation with Jupiter creates bulges on Io just like the Moon creates the ocean tides on Earth. The constant change in size and orientation of Io causes friction that creates enough internal heat for volcanic eruptions to occur.

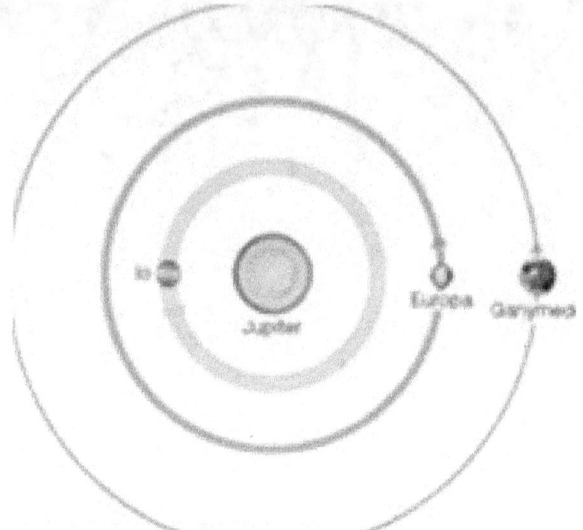

IO'S ELLIPTICAL ORBIT

Io's Elliptical Orbit

Ganymede, Europa and Io are all in orbital resonance with Jupiter. Io completes exactly four orbits and Europa completes exactly two orbits in the same time it takes Ganymede to complete one orbit around Jupiter. During the course of their orbits, the three moons line up like in the picture seen above. Since they periodically line up in this fashion, the gravitational tugs the moons exert on each other stretch their orbits into elliptical shapes.

EUROPA

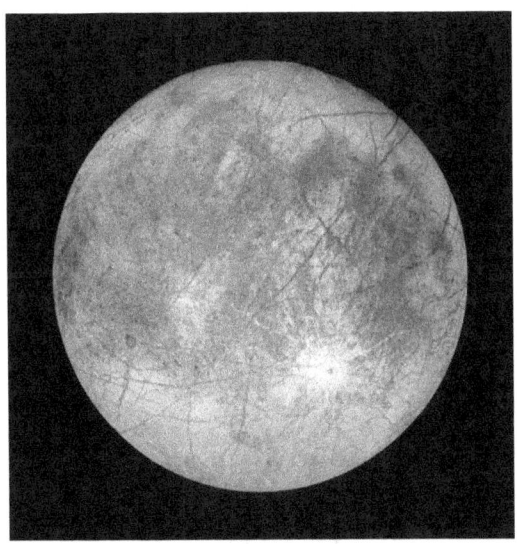

This moon of Jupiter might have an Ocean under the ice. Scientists have already proposed missions to send an ocean exploring robot there.

Robotic Submarine on Europa

EUROPA: WHAT LIES BENEATH?

Europa is thought to be the most likely place in the Solar System to harbor life since it seems to have a large water ocean covered by ice. Life on Europa is a theme of many movies including the movie "2010".

Europa's surface and crust are made almost entirely of water ice, and its bizarre, fractured appearance is proof enough that tidal heating has acted there. The icy surface is nearly devoid of impact craters and may be only a few million years old.

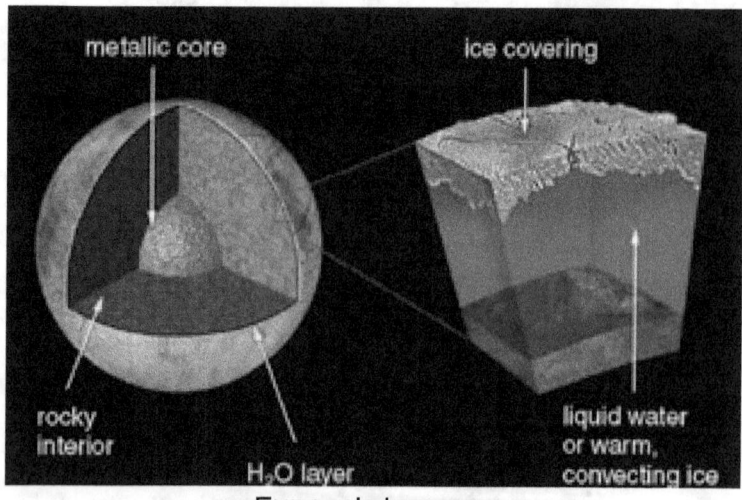

EUROPA'S INTERIOR

Observations made by the Galileo spacecraft show that Europa has a metallic core and a rocky mantle.

Surrounding the rocky interior appears to be an icy layer 100 kilometers thick, the top few kilometers of which seem to be frozen solid. The stretching and squeezing of tidal friction should provide enough heat to melt some of this into liquid water beneath a thin ice shell. If it does, then Europa may have an ocean with more than twice as much liquid water as all of Earth's oceans combined.

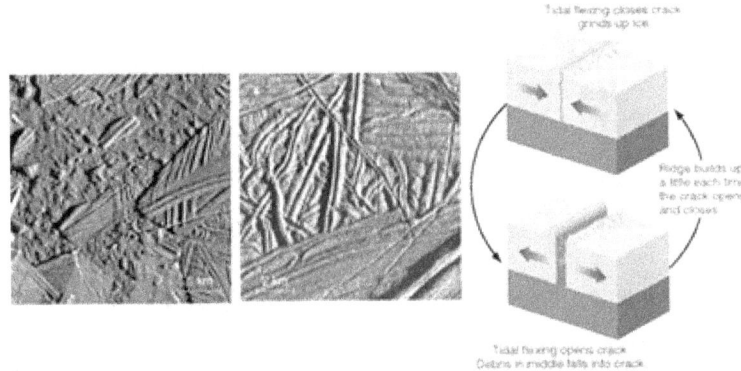

ANALYZING EUROPA'S CRACKED SURFACE

Close-up photos of the surface of Europa support the idea of a liquid ocean beneath the surface. These photos, taken by the Galileo spacecraft, show what appears to be icebergs stuck in a layer of ice. Other evidence comes from double-ridged cracks on the surface. Tidal flexing that allows water to well up and build ridges may create these cracks.

GANYMEDE:
LARGEST MOON IN THE SOLAR SYSTEM

Ganymede is the largest moon in the solar system and has the second highest moon gravity. It would make a great place to settle and build a base in the Jovian system.

A Possible Base on Ganymede

Two Regions of Ganymede's Surface

The surface of Ganymede shares many similarities with Europa. Ganymede's surface is also made of water ice, but unlike Europa's surface, it shows signs of varying age. The darker regions are heavily cratered, suggesting they are billions of years old. The lighter regions show no signs of craters and it is thought that eruptions of water covered the surface before freezing over. These areas are geologically younger than the darker regions.

If liquid water occasionally makes its way to the surface to fill in craters, could that suggest a liquid ocean similar to the one that might exist on Europa?

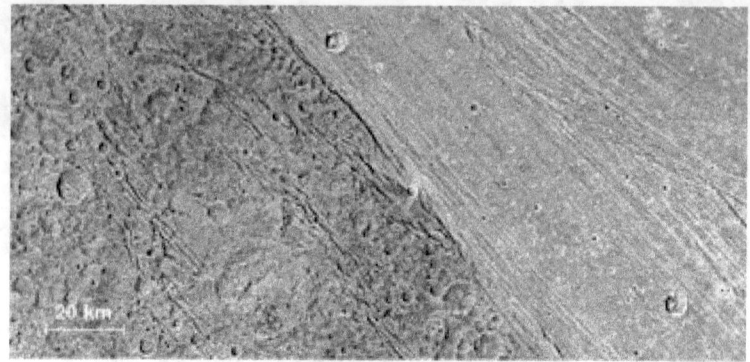

GANYMEDE'S SURFACE CLOSE-UP

Not necessarily. The case for Europa's subsurface ocean comes from the strong probability of tidal heating, melting the ice under the surface. Ganymede has a much weaker tidal force, and thus weaker tidal heating than Europa and Io. The level of Ganymede's tidal heating could not provide enough heat to make an ocean of liquid water. Aside from tidal heating, we are not sure where sufficient heat would come from to melt the ice.

CALLISTO

THE OUTERMOST GALILEAN MOON

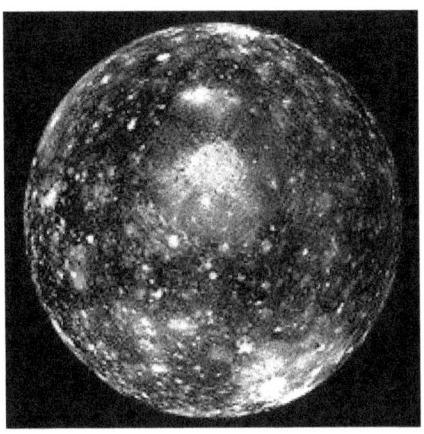

CALLISTO

Callisto is the stereotypical outer solar system satellite. It is one of the largest and most heavily cratered satellites in the solar system. The surface is very icy and dates back four billion years. Beneath the icy crust is possibly a salty ocean supported by a deeper rocky interior.

Callisto's Surface

Callisto doesn't have any large mountains, show evidence of volcanic or tectonic activity or have any appreciable level of internal heat. Nonetheless, observations of Callisto's magnetic field may cause scientists to add the large moon to the list of possible worlds with subsurface salty oceans.

There are a total of 79 Moons of Jupiter. Some of the other most interesting (besides the Galilean Satellites) are described below:

Metis

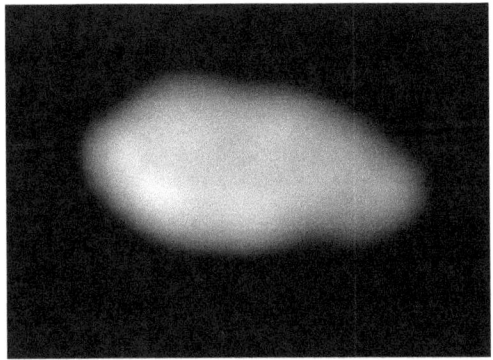

Discovered by Stephen Synnott through images taken by the Voyager 1 in 1979, Metis is the closest moon to Jupiter. It has a diameter of 40 km (25 miles) and orbits Jupiter in 0.294780 Earth days, which is faster than Jupiter rotates on its axis.

It is named after the Greek Titaness Metis, who was the first wife of the god of the skies, Zeus. It is thought that Metis is an asteroid that was captured by Jupiter's gravity. The moon orbits at 128,000 km (79,500 miles) from Jupiter and has a mass of 9×10^{16} kg.

Adrastea

The second closest moon to Jupiter is Adrastea which has a diameter of 20 km (12 miles) and orbits 129,000 km (80,000 miles) from Jupiter. It was discovered in 1979 by David Jewitt using the Voyager 2 and is named for the Greek goddess Adrasteia, who passed out rewards and punishments.

Adrastea has a mass of 1.91×10^{16} kg and like Metis it orbits Jupiter faster than it takes for Jupiter to rotate on its axis – in 0.29826 Earth days.

Amalthea

Amalthea is the third moon from Jupiter and is the reddest object in the solar system. It was discovered by Edward Emerson Barnard in 1892. Amalthea is not a spherical moon with a diameter of 232 x 146 x 134 km (145 x 91 x 83 miles). It orbits 181,300 km (112,700 miles) from Jupiter, located in the faint Gossamer ring.

The moon is named for Amalthea in Greek mythology, who was the foster-mother of Zeus and nursed the god of the skies when he was a baby with goats milk. It is the largest of the inner moons of Jupiter and is most likely an asteroid captured by Jupiter. It takes Amalthea 0.49817905 Earth days to orbit Jupiter and like all the inner moons, it is tidally locked to the planet – the same side of the moon always faces Jupiter. Amalthea also gives off more heat than it receives from the Sun.

Thebe

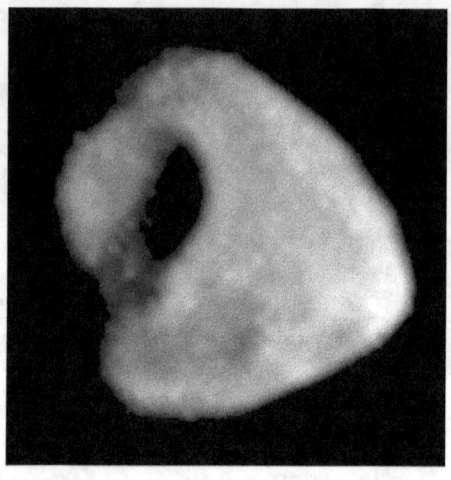

The fourth moon from Jupiter is Thebe. It has a diameter of 110 x 90 km (68 x 56 miles) and orbits 222,000 km (138,000 miles) from Jupiter. Thebe was discovered by Stephen P. Synnott in 1979 and officially named in 1983.

In Greek mythology, Thebe was a nymph and the daughter of the river god Asopus. It is likely that Thebe, along with Amalthea, provide the dust for the Gossamer ring where they are located. Thebe has a mass of 8 x 10^{17}kg and it takes the moon 0.6745 Earth days to orbit Jupiter.

Leda

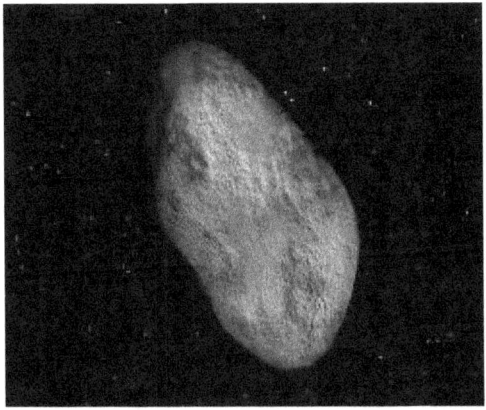

Leda is the ninth moon from Jupiter and is also the smallest moon with a mean diameter of 16 km (9.9 miles). Charles Kowal discovered Leda in 1974. It is named for the queen of Sparta and the mother of Pollux and Helen of Troy – the father was Zeus. The moon has a mass of 5.68 x 10^{15}kg. It takes 238.72 Earth days for Leda to orbit Jupiter and it orbits at a distance of 11,094,000 km (6,900,000 miles) from the planet.

Himalia

Jupiter's tenth moon is Himalia, discovered by Charles Perrine in 1904. Himalia is 170 km (110 miles) in diameter and orbits 11,480,000 km (7,000,000 miles) from Jupiter. The moon is named after a nymph who produced three sons with Zeus (Jupiter). It has a mass of 9.5 x 10^{18}kg and takes 250.5662 days to orbit Jupiter.

Lysithea

The eleventh moon from Jupiter's surface is Lysithea, a world which is 24 km (15 miles) in diameter and orbits at around 11,720,000 km (7,200,000 miles) from Jupiter. It has a mass of 8 x 10^{16}kg and takes 259.22 Earth days to orbit Jupiter. Lysithea is named after the daughter of Oceanus in Greek mythology. Lysithea was discovered in 1938 by Seth Nicholson.

Elara

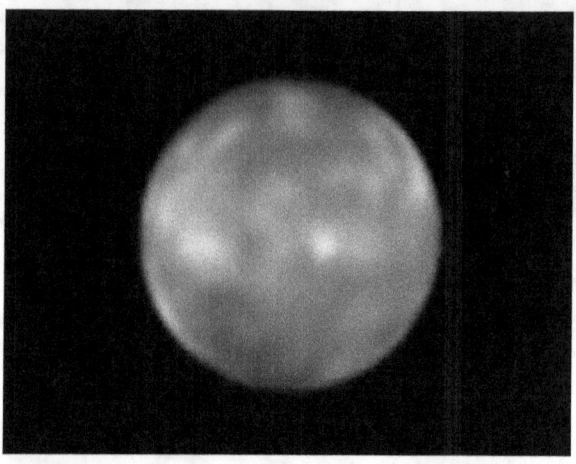

Elara was discovered in 1905 by Charles Perinne and is Jupiter's twelfth moon. It has a diameter of 80 km (50 miles) and orbits Jupiter at a distance of 11,737,000 km (7,250,000 miles). Elara has a mass of 8 x 10^{17}kg and takes 259.6528 Earth days to orbit Jupiter. It is named after Elara from Greek mythology, the mother of the giant Tityus, fathered by Zeus.

Ananke

Ananke, Jupiter's thirteenth moon, was discovered in 1951 by Seth Nicholson. Ananke has a diameter of 20 km (12.5 miles) and orbits 21,200,000 km (13,100,000 miles) from Jupiter. The moon has a mass of 4×10^{16}kg and it take Ananke 631 Earth days to orbit Jupiter. The moon is also in a retrograde orbit – which means it orbits in the opposite direction of Jupiter. It is named after Ananke, the mother of Adrastea by Zeus, in Greek mythology.

Carme

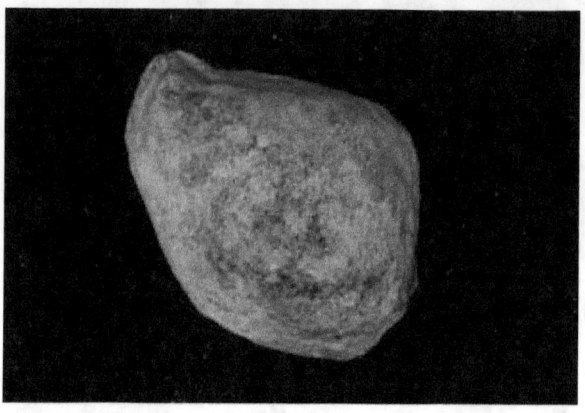

Discovered in 1938 by Charles Nicholson, Carme is the fourteenth moon of Jupiter. It has a diameter of 30 km (18.5 miles) and orbits at a distance of 22,600,000 km (13,800,000 miles) from Jupiter. Carme has a mass of 9 x 10^{16}kg and orbits Jupiter in 692 Earth days. It is in a retrograde orbit which moves in the opposite direction of Jupiter. In Greek mythology Carme was the mother of Britomartis, a Cretan goddess, fathered by Zeus.

Pasiphae

Pasiphae is the fifteenth moon of Jupiter and was discovered by P. Melotte in 1908. It orbits Jupiter at a distance of 23,500,000 km (14,600,000 miles) and has a diameter of 36 km (22 miles). Its mass is 2 x 10^{23}kg and it takes Pasiphae 735 Earth days to orbit Jupiter in a retrograde orbit path. In Greek mythology, Pasiphae was the wife of Minos and mother of the Minotaur.

Sinope

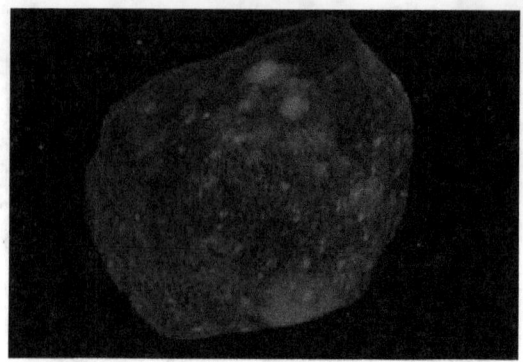

Jupiter's sixteenth moon is Sinope, discovered in 1914 by Seth Nicholson. Sinope has a diameter of 28 km (17.5 miles) and it orbits Jupiter at a distance of 23,700,000 km (14,700,000 miles). It has a mass of 8×10^{16} kg and it orbits Jupiter in a retrograde orbit that takes 758 Earth days. In Greek mythology, Sinope was a woman who was courted unsuccessfully by Zeus, and she remained a virgin for her entire life.

Callirrhoe

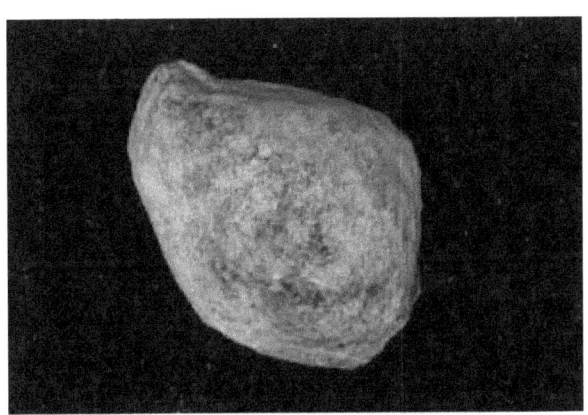

Jupiter's seventeenth confirmed moon was Callirrhoe, also known as S/1999 J 1, and was discovered by Tim Spahr on July 18, 2000. It has a diameter of 8.6 km (5.3 miles) and orbits Jupiter at a distance of 24,100,000 km (14,975,000 miles). Callirrhoe has a mass of 9×10^{14} kg and orbits Jupiter in a retrograde orbit that takes 758.77 Earth days to complete. In Greek mythology, Callirrhoe was the daughter of the river god Achelous, one of Zeus' (Jupiter's) many conquests.

3.4 The Moons of Saturn

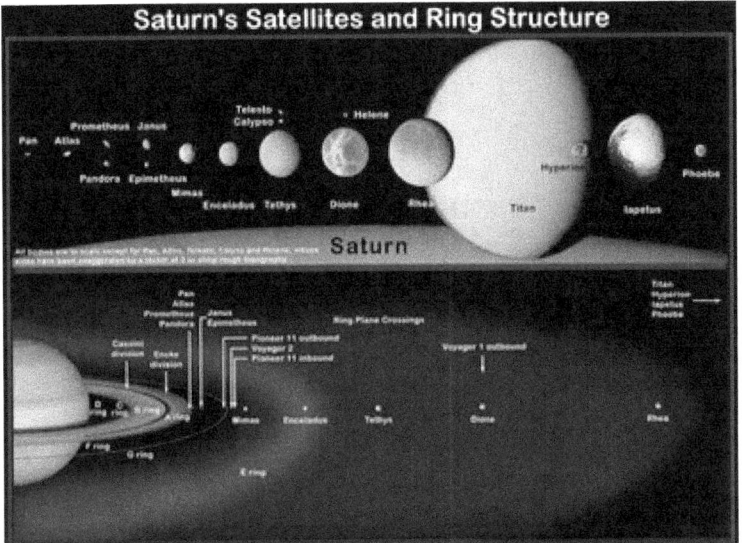

The moons of Saturn are numerous and diverse, ranging from tiny moonlets only tens of meters across to enormous Titan, which is larger than the planet Mercury. Saturn has 82 moons with confirmed orbits that are not embedded in its rings – of which only 13 have diameters greater than 50 kilometers – as well as dense rings that contain millions of embedded moonlets and innumerable smaller ring particles. Seven Saturnian moons are large enough to have collapsed into a relaxed, ellipsoidal shape, though only one or two of those, Titan and possibly Rhea, are currently in hydrostatic equilibrium.

Particularly notable among Saturn's moons are Titan, the second-largest moon in the Solar System (after Jupiter's Ganymede), with a nitrogen-rich Earth-like atmosphere and a landscape featuring dry river networks and hydrocarbon lakes, Enceladus, which emits jets of gas and dust from its south-polar region, and Iapetus, with its contrasting black and white hemispheres.

Some of the major interesting moons of Saturn include these:

Titan

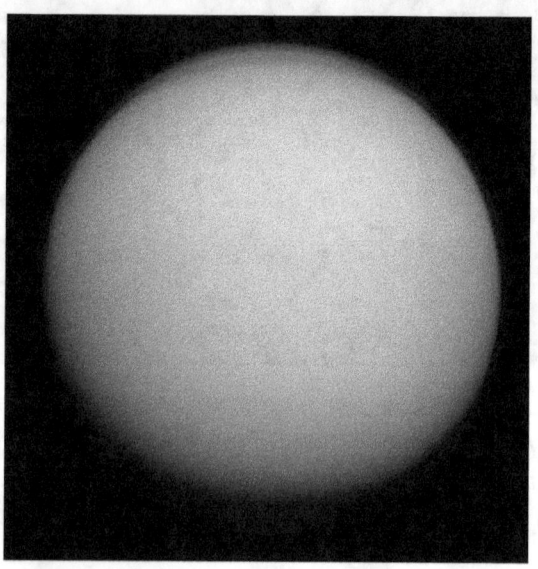

Titan would be the logical place to build a base in the Saturnian System. It has an atmosphere (even though it is very cold a deadly) and so would allow airplanes for travel. Its surface would be a good place to build bases.

A Base Concept on Titan

Titan is the largest moon of Saturn and the second-largest natural satellite in the Solar System. It is the only moon known to have a dense atmosphere, and the only known body in space, other than Earth, where clear evidence of stable bodies of surface liquid has been found.

Titan is one of seven gravitationally rounded moons in orbit around Saturn, and the second most distant from Saturn of those seven. Frequently described as a planet-like moon, Titan is 50% larger (in diameter) than Earth's Moon and 80% more massive. It is the second-largest moon in the Solar System after Jupiter's moon Ganymede, and is larger than the planet Mercury, but only 40% as massive.

Discovered in 1655 by the Dutch astronomer Christiaan Huygens, Titan was the first known moon of Saturn, and the sixth known planetary satellite (after Earth's moon and the four Galilean moons of Jupiter). Titan orbits Saturn at 20 Saturn radii. From Titan's surface, Saturn subtends an arc of 5.09 degrees and, were it visible through the moon's thick atmosphere, would appear 11.4 times larger in the sky than the Moon from Earth.

The Cassini Hugens Mission

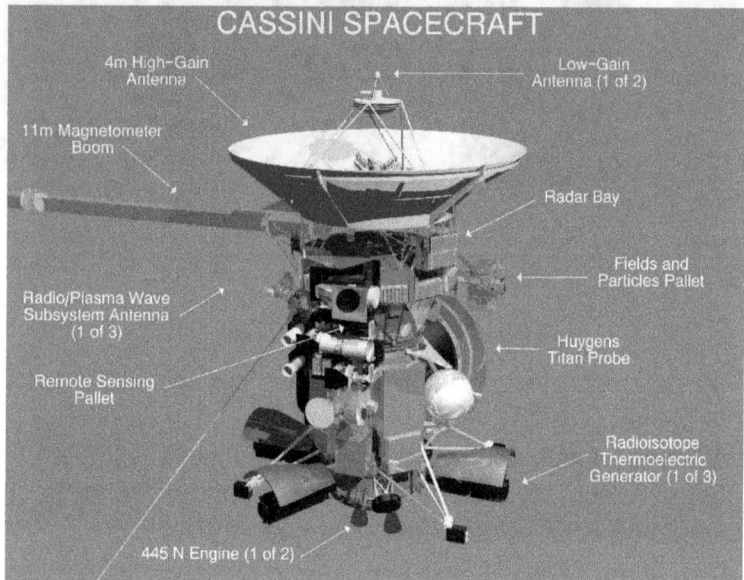

CASSINI SPACECRAFT

- 4m High-Gain Antenna
- Low-Gain Antenna (1 of 2)
- 11m Magnetometer Boom
- Radar Bay
- Fields and Particles Pallet
- Radio/Plasma Wave Subsystem Antenna (1 of 3)
- Huygens Titan Probe
- Remote Sensing Pallet
- Radioisotope Thermoelectric Generator (1 of 3)
- 445 N Engine (1 of 2)

The Cassini–Huygens spacecraft reached Saturn on July 1, 2004, and began the process of mapping Titan's surface by radar. A joint project of the European Space Agency (ESA) and NASA, Cassini–Huygens proved a very successful mission. The Cassini probe flew by Titan on October 26, 2004, and took the highest-resolution images ever of Titan's surface, at only 1,200 kilometers (750 mi), discerning patches of light and dark that would be invisible to the human eye.

On July 22, 2006, Cassini made its first targeted, close fly-by at 950 kilometers (590 mi) from Titan; the closest flyby was at 880 kilometers (550 mi) on June 21, 2010. Liquid has been found in abundance on the surface in the North Polar Region, in the form of many lakes and seas discovered by Cassini.

Dione

Dione is thought to be a dense rocky core surrounded by water-ice. The tidally locked moon is heavily cratered not on its leading side but on its back side. Astronomers think a collision could have spun the moon on its axis. The moon hosts a thin oxygen atmosphere and may have a liquid ocean beneath its surface.

At 1122 km (697 mi) in diameter, Dione is the 15th largest moon in the Solar System, and is more massive than all known moons smaller than itself combined. About two thirds of Dione's mass is water ice, and the remaining is a dense core, probably silicate rock.

Enceladus

Cassini revealed the dramatic truth: Enceladus is an active moon that hides a global ocean of liquid salty water beneath its crust. What's more, jets of icy particles from that ocean, laced with a brew of water and simple organic chemicals, gush out into space continuously from this fascinating ocean world. It is a great potential site for other life existing in our system.

Enceladus contains more than 100 geysers at its south pole. Tidal heating causes portions of the icy planet to melt, spewing icy material into space from its "tiger stripes." The tiny bits of ice travel together to create Saturn's E ring. The satellite's icy surface makes it one of the brightest objects in the solar system. The moon has a subsurface ocean that may be friendly to life.

Hyperion

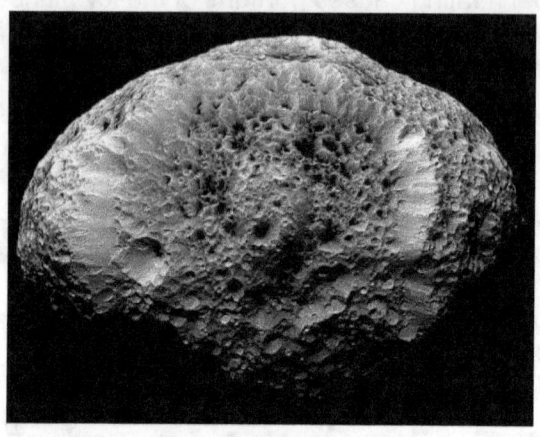

Hyperion was the last of the major satellites to be discovered. Hyperion is a small moon with an irregular appearance. The flattened object resembles an elongated potato rather than a sphere, a form that may have been created when an impact demolished a larger moon long ago. Hyperion has a spongy shape, possibly due to its low density and porous surface. Impacts seem to be absorbed by the moon, and most of the ejecta is thrown into space.

Iapetus

Iapetus features light and dark contrasts on its surface, giving the moon a yin-yang shape. Dark hydrocarbons falling to the moon long ago, perhaps from the nearby moon Phoebe, may have had more time to absorb more heat, gradually growing and spreading over time. Iapetus has a walnut-like shape, with its center bulging outward, and a ridge running around its equator. The moon also contains some of the highest mountains in the solar system, which may have been material from another moon. Scientists are studying ice movements (such as landslides) to do comparative work with these types of features on Earth.

The equatorial ridge of Iapetus can reach heights of up to 12 miles (20 km). This image reveals mountains only about half that height.

Mimas

Mimas seems to be made from a lot of if not all ice and that would make it also a great place for a colony or at least to do ice mining for the rest of the Solar System.

Mimas has a gaping crater that gives the rocky moon a strong resemblance to fictional Death Star in the "Star Wars" movies. The impact stands out despite the fact that Mimas is one of the most heavily cratered bodies in the solar system, with overlapping impacts covering the surface. The smallest and closest orbiting of Saturn's major moons, Mimas cleared the gap known as the Cassini division between two of the planet's rings. Mimas is made up primarily of water-ice, but despite its proximity to the planet (and the resulting tidal heating that should occur), the surface of the moon remains unchanged; none of the ice seems to be melting, though such melting occurs on other, more distant moons. It is possible that there is a liquid ocean beneath its surface, although scientists say that an oval-shaped core could also explain some of Mimas' libration movements.

Rhea

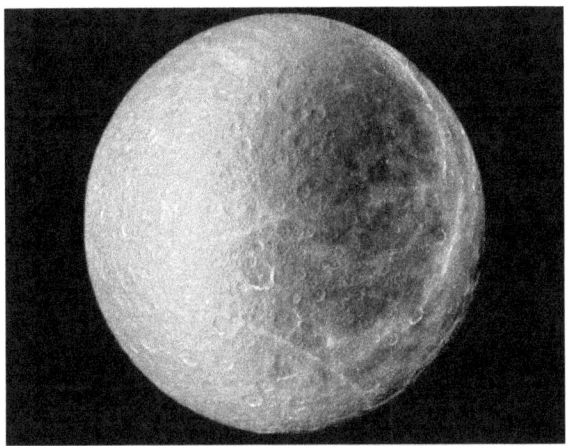

(Rhea in False Color)

Rhea is a heavily cratered moon and lacks a core at its center. Instead, the entire body is composed of ice, with traces of rock mixed in, causing it to resemble a dirty snowball. The second largest of the planet's major moons, Rhea is still rather small, about half the size of Earth's moon. The satellite contains a faint oxygen atmosphere, about 5 trillion times less dense than the one found on Earth, but the only known oxygen atmosphere in the solar system. Radiation from Saturn's magnetosphere could release oxygen and carbon dioxide from the icy surface.

Tethys

Tethys travels close to Saturn and feels the gravitational pull of the planet. The heat from Saturn may allow the moon's icy surface to melt slightly, filling in craters and other signs of impact. Made up almost entirely of water ice, the surface is highly reflective. A large trench crosses the moon, running diagonally from its north to South Pole and spanning three-quarters of the satellite's circumference. A large crater on the other side of the moon covers nearly two-fifth of the moon's diameter and is nearly the size of Mimas. Scientists have found strange red arcs on the moon and are still struggling to explain how the arcs got there.

3.5 The Moons of Neptune

Triton

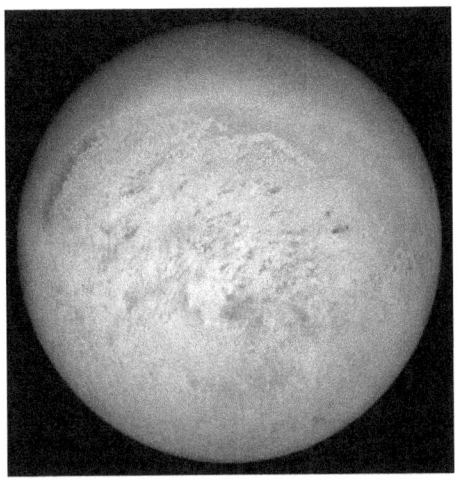

Triton follows a retrograde and quasi-circular orbit, and is thought to be a gravitationally captured satellite. It was the second moon in the Solar System that was discovered to have a substantial atmosphere, which is primarily nitrogen with small amounts of methane and carbon monoxide.

The pressure on Triton's surface is about 14 μbar. In 1989 the Voyager 2 spacecraft observed what appeared to be clouds and hazes in this thin atmosphere. Triton is one of the coldest bodies in the Solar System, with a surface temperature of about 38 K (−235.2 °C). Its surface is covered by nitrogen, methane, carbon dioxide and water ices and has a high geometric albedo of more than 70%. The Bond albedo is even higher, reaching up to 90%.Surface features include the large southern polar cap, older cratered planes cross-cut by graben and scarps, as well as youthful features probably formed by endogenic processes like cryovolcanism. Voyager 2 observations revealed a number of active geysers within the polar cap heated by the Sun, which eject plumes to the height of up to 8 km. Triton has a relatively high density of about 2 g/cm3 indicating that rocks constitute about two thirds of its mass, and ices (mainly water ice) the remaining one third. There may be a layer of liquid water deep inside Triton, forming a subterranean ocean. Because of its retrograde orbit and relative proximity to Neptune (closer than the Moon is to Earth), tidal deceleration is causing Triton to spiral inward, which will lead to its destruction in about 3.6 billion years.

Nereid

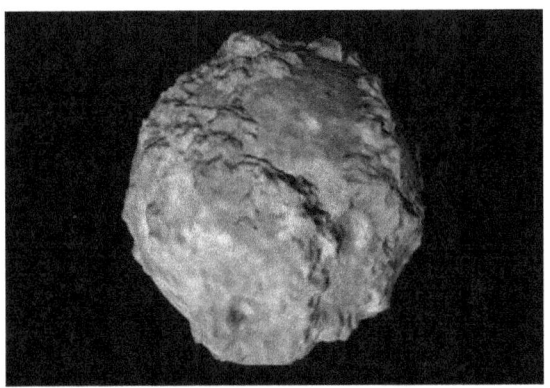

Nereid is the third-largest moon of Neptune. It has a prograde but very eccentric orbit and is believed to be a former regular satellite that was scattered to its current orbit through gravitational interactions during Triton's capture. Water ice has been spectroscopically detected on its surface. Early measurements of Nereid showed large, irregular variations in its visible magnitude, which were speculated to be caused by forced precession or chaotic rotation combined with an elongated shape and bright or dark spots on the surface. This was disproved in 2016, when observations from the Kepler space telescope showed only minor variations. Thermal modeling based on infrared observations from the Spitzer and Herschel space telescopes suggest that Nereid is only moderately elongated which disfavours forced precession of the rotation. The thermal model also indicates that the surface roughness of Nereid is very high, likely similar to the Saturnian moon Hyperion.

3.6 The Moons of Uranus

Uranus, the seventh planet of the Solar System, has 27 known moons, most of which are named after characters that appear in, or are mentioned in, the works of William Shakespeare and Alexander Pope. Uranus's moons are divided into three groups: thirteen inner moons, five major moons, and nine irregular moons. The inner and major moons all have prograde orbits, while orbits of the irregulars are mostly retrograde. The inner moons are small dark bodies that share common properties and origins with Uranus's rings. The five major moons are ellipsoidal, indicating that they reached hydrostatic equilibrium at some point in their past (and may still be in equilibrium), and four of them show signs of internally driven processes such as canyon formation and volcanism on their surfaces.

The largest of these five, Titania, is 1,578 km in diameter and the eighth-largest moon in the Solar System, about one-twentieth the mass of the Earth's Moon. The orbits of the regular moons are nearly coplanar with Uranus's equator, which is tilted 97.77° to its orbit. Uranus's irregular moons have elliptical and strongly inclined (mostly retrograde) orbits at large distances from the planet.

William Herschel discovered the first two moons, Titania and Oberon, in 1787. The other three ellipsoidal moons were discovered in 1851 by William Lassell (Ariel and Umbriel) and in 1948 by Gerard Kuiper (Miranda). These five have planetary mass, and so would be considered (dwarf) planets if they were in direct orbit about the Sun. The remaining moons were discovered after 1985, either during the Voyager 2 flyby mission or with the aid of advanced Earth-based telescopes.

Ariel

Discovered on 24 October 1851 by William Lassell, it is named for a sky spirit in Alexander Pope's The Rape of the Lock and Shakespeare's The Tempest.

After Miranda, Ariel is the second-smallest of Uranus' five major rounded satellites and the second-closest to its planet. Among the smallest of the Solar System's 19 known spherical moons (it ranks 14th among them in diameter), it is believed to be composed of roughly equal parts ice and rocky material. Its mass is approximately equal in magnitude to Earth's hydrosphere.

Like all of Uranus' moons, Ariel probably formed from an accretion disc that surrounded the planet shortly after its formation, and, like other large moons, it is likely differentiated, with an inner core of rock surrounded by a mantle of ice. Ariel has a complex surface consisting of extensive cratered terrain cross-cut by a system of scarps, canyons, and ridges. The surface shows signs of more recent geological activity than other Uranian moons, most likely due to tidal heating.

Miranda

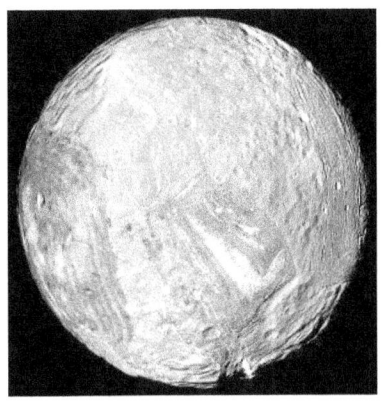

Miranda, also designated Uranus V, is the smallest and innermost of Uranus's five round satellites. It was discovered by Gerard Kuiper on 16 February 1948 at McDonald Observatory in Texas, and named after Miranda from William Shakespeare's play The Tempest. Like the other large moons of Uranus, Miranda orbits close to its planet's equatorial plane. Because Uranus orbits the Sun on its side, Miranda's orbit is perpendicular to the ecliptic and shares Uranus' extreme seasonal cycle.

At just 470 km in diameter, Miranda is one of the smallest closely observed objects in the Solar System that might be in hydrostatic equilibrium (spherical under its own gravity). The only close-up images of Miranda are from the Voyager 2 probe, which made observations of Miranda during its Uranus flyby in January 1986. During the flyby, Miranda's southern hemisphere pointed towards the Sun, so only that part was studied.

Titania

Titania also designated Uranus III, is the largest of the moons of Uranus and the eighth largest moon in the Solar System at a diameter of 1,578 kilometers (981 mi). Discovered by William Herschel in 1787, Titania is named after the queen of the fairies in Shakespeare's A Midsummer Night's Dream. Its orbit lies inside Uranus's magnetosphere.

Titania consists of approximately equal amounts of ice and rock, and is probably differentiated into a rocky core and an icy mantle. A layer of liquid water may be present at the core–mantle boundary. The surface of Titania, which is relatively dark and slightly red in color, appears to have been shaped by both impacts and endogenic processes. It is covered with numerous impact craters reaching up to 326 kilometers (203 mi) in diameter, but is less heavily cratered than Oberon, outermost of the five large moons of Uranus. Titania probably underwent an early endogenic resurfacing event which obliterated its older, heavily cratered surface. Titania's surface is cut by a system of enormous canyons and scarps, the result of the expansion of its interior during the later stages of its evolution. Like all major moons of Uranus, Titania probably formed from an accretion disk which surrounded the planet just after its formation.

Infrared spectroscopy conducted from 2001 to 2005 revealed the presence of water ice as well as frozen carbon dioxide on the surface of Titania, which in turn suggested that the moon may have a tenuous carbon dioxide atmosphere with a surface pressure of about 10 nanopascals (10−13 bar). Measurements during Titania's occultation of a star put an upper limit on the surface pressure of any possible atmosphere at 1–2 mPa (10–20 nbar).

Oberon

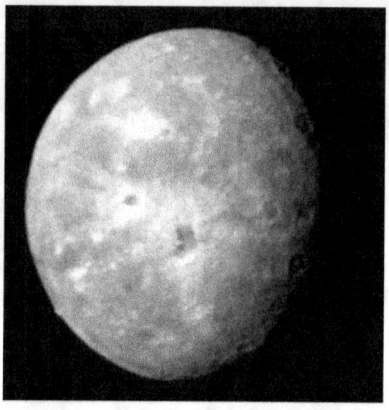

Oberon, also designated Uranus IV, is the outermost major moon of the planet Uranus. It is the second-largest and second most massive of the Uranian moons, and the ninth most massive moon in the Solar System. Discovered by William Herschel in 1787, Oberon is named after the mythical king of the fairies who appears as a character in Shakespeare's A Midsummer Night's Dream. Its orbit lies partially outside Uranus's magnetosphere.

It is likely that Oberon formed from the accretion disk that surrounded Uranus just after the planet's formation. The moon consists of approximately equal amounts of ice and rock, and is probably differentiated into a rocky core and an icy mantle. A layer of liquid water may be present at the boundary between the mantle and the core. The surface of Oberon, which is dark and slightly red in color, appears to have been primarily shaped by asteroid and comet impacts. It is covered by numerous impact craters reaching 210 km in diameter. Oberon possesses a system of chasmata (graben or scarps) formed during crustal extension as a result of the expansion of its interior during its early evolution.

Umbriel

Umbriel is a moon of Uranus discovered on October 24, 1851, by William Lassell. It was discovered at the same time as Ariel and named after a character in Alexander Pope's poem The Rape of the Lock. Umbriel consists mainly of ice with a substantial fraction of rock, and may be differentiated into a rocky core and an icy mantle. The surface is the darkest among Uranian moons, and appears to have been shaped primarily by impacts. However, the presence of canyons suggests early endogenic processes, and the moon may have undergone an early endogenically driven resurfacing event that obliterated its older surface.

Covered by numerous impact craters reaching 210 km (130 mi) in diameter, Umbriel is the second most heavily cratered satellite of Uranus after Oberon. The most prominent surface feature is a ring of bright material on the floor of Wunda crater. This moon, like all moons of Uranus, probably formed from an accretion disk that surrounded the planet just after its formation

Exploring and Settling Our Huge Solar System

3.7 The Moons of Pluto

The innermost and largest moon, Charon, was discovered by James Christy on 22 June 1978, nearly half a century after Pluto was discovered. This led to a substantial revision in estimates of Pluto's size, which had previously assumed that the observed mass and reflected light of the system were all attributable to Pluto alone.

Two additional moons were imaged by astronomers of the Pluto Companion Search Team preparing for the New Horizons mission and working with the Hubble Space Telescope on 15 May 2005, which received the provisional designations S/2005 P 1 and S/2005 P 2. The International Astronomical Union officially named these moons Nix (or Pluto II, the inner of the two moons, formerly P 2) and Hydra (Pluto III, the outer moon, formerly P 1), on 21 June 2006. Kerberos, announced on 20 July 2011, was discovered while searching for Plutonian rings. Styx, announced on 7 July 2012, was discovered while looking for potential hazards for New Horizons.

Charon

Charon also known as (134340) Pluto I, is the largest of the five known natural satellites of the dwarf planet Pluto. It has a mean radius of 606 km (377 mi). Charon is the sixth-largest trans-Neptunian object after Pluto, Eris, Haumea, Makemake and Gonggong. It was discovered in 1978 at the United States Naval Observatory in Washington, D.C., using photographic plates taken at the United States Naval Observatory Flagstaff Station (NOFS).

With half the diameter and one eighth the mass of Pluto, Charon is a very large moon in comparison to its parent body. Its gravitational influence is such that the barycenter of the Plutonian system lies outside Pluto.

The reddish-brown cap of the north pole of Charon is composed of tholins, organic macromolecules that may be essential ingredients of life. These tholins were produced from methane, nitrogen and related gases released from the atmosphere of Pluto and transferred over 19,000 km (12,000 mi) to the orbiting moon.

Hydra

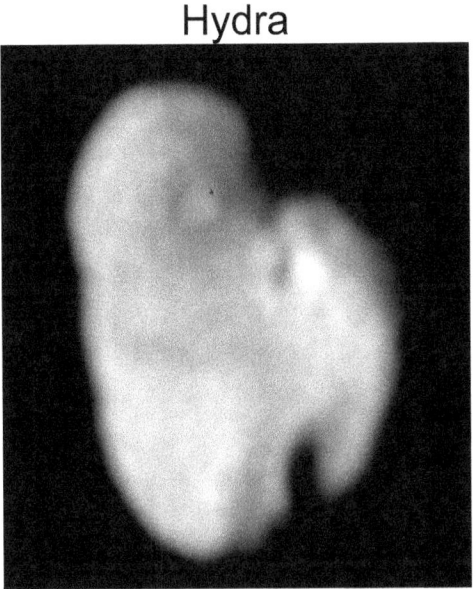

Hydra is a natural satellite of Pluto, with a diameter of approximately 51 km (32 mi) across its longest dimension. It is the second largest moon of Pluto, being slightly larger than Nix. Hydra was discovered along with Nix by the Pluto Companion Search Team in June 2005. It was named after the Hydra, the nine-headed underworld serpent in Greek mythology. By distance, Hydra is the fifth and outermost moon of Pluto, orbiting beyond Pluto's fourth moon Kerberos.

Hydra has a highly reflective surface caused by the presence of water ice, similar to other plutonian moons. Hydra's reflectivity is intermediate, in between those of Pluto and Charon.

4.0 Asteroids of the Solar System

4.1 The Major Asteroids

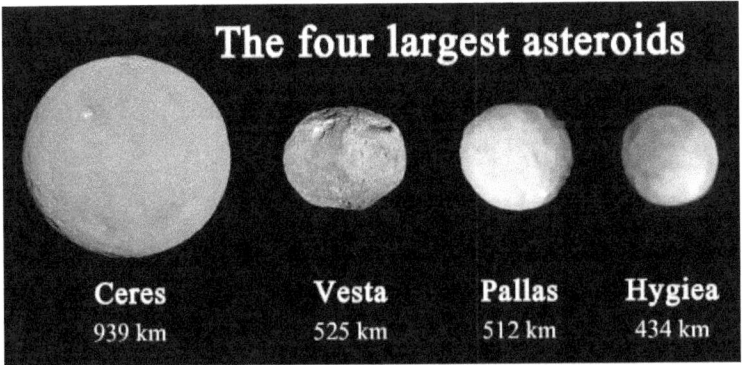

In this chapter we will show some of the exceptional asteroids in the Solar System. For the purposes of clarification "asteroid" means minor planet up to the orbit of Jupiter, which includes the dwarf planet Ceres.

Asteroids are given minor planet numbers, but not all minor planets are asteroids. Minor planet numbers are also given to objects of the Kuiper belt, which is similar to the asteroid belt but farther out around 30–60 AU, whereas asteroids are mostly between 2–3 AU from the Sun. Also, comets are not typically included under minor planet numbers, and have their own naming conventions.

Asteroids are given a unique sequential identifying number once their orbit is precisely determined. Prior to this, they are known only by their systematic name or provisional designation, such as 1950 DA.

4.2 Trojans & Greeks

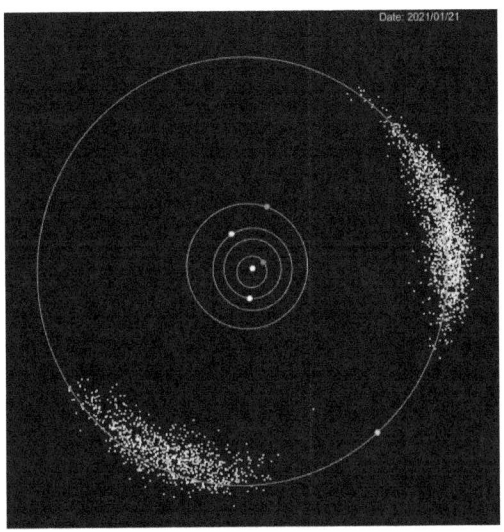

The Jupiter Trojans, commonly called Trojan asteroids or simply Trojans, are a large group of asteroids that share the planet Jupiter's orbit around the Sun. Relative to Jupiter, each Trojan librates around one of Jupiter's stable Lagrange points: either L4, lying 60° ahead of the planet in its orbit, or L5, 60° behind. Jupiter Trojans are distributed in two elongated, curved regions around these Lagrangian points with an average semi-major axis of about 5.2 AU.

The first Jupiter Trojan discovered, 588 Achilles, was spotted in 1906 by German astronomer Max Wolf. A total of 7,040 Jupiter Trojans have been found as of October 2018. By convention, they are each named from Greek mythology after a figure of the Trojan War, hence the name "Trojan". The total number of Jupiter Trojans larger than 1 km in diameter is believed to be about 1 million, approximately equal to the number of asteroids larger than

1 km in the asteroid belt. Like main-belt asteroids, Jupiter Trojans form families.

As of 2004, many Jupiter Trojans showed to observational instruments as dark bodies with reddish, featureless spectra. No firm evidence of the presence of water, or any other specific compound on their surface has been obtained, but it is thought that they are coated in tholins, organic polymers formed by the Sun's radiation. The Jupiter Trojans' densities (as measured by studying binaries or rotational lightcurves) vary from 0.8 to 2.5 g·cm−3. Jupiter Trojans are thought to have been captured into their orbits during the early stages of the Solar System's formation or slightly later, during the migration of giant planets.

The term "Trojan Asteroid" specifically refers to the asteroids co-orbital with Jupiter, but the general term "Trojan" is sometimes more generally applied to other small Solar System bodies with similar relationships to larger bodies: for example, there are both Mars Trojans and Neptune Trojans, as well as a recently discovered Earth Trojan. The term "Trojan asteroid" is normally understood to specifically mean the Jupiter Trojans because the first Trojans were discovered near Jupiter's orbit and Jupiter currently has by far the most known Trojans.

4.3 Near Earth Asteroids

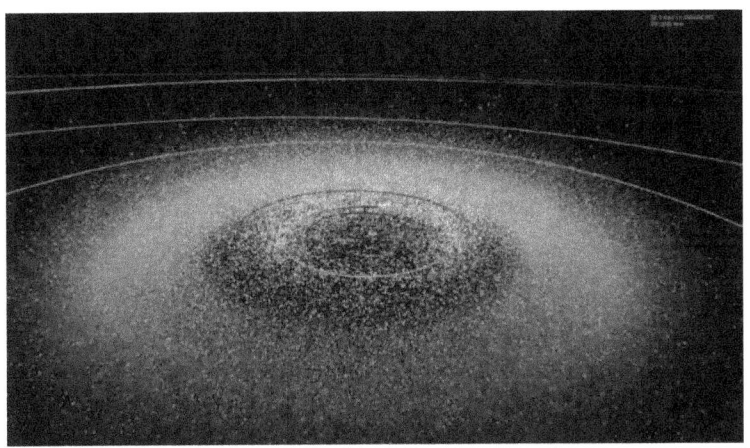

These are objects in a near-Earth orbit without the tail or coma of a comet. As of March 5, 2020, 22,261 near-Earth asteroids are known, 1,955 of which are both sufficiently large and come sufficiently close to Earth to be considered potentially hazardous.

NEAs survive in their orbits for just a few million years. They are eventually eliminated by planetary perturbations, causing ejection from the Solar System or a collision with the Sun or a planet. With orbital lifetimes short compared to the age of the Solar System, new asteroids must be constantly moved into near-Earth orbits to explain the observed asteroids. The accepted origin of these asteroids is that main-belt asteroids are moved into the inner Solar System through orbital resonances with Jupiter.

The interaction with Jupiter through the resonance perturbs the asteroid's orbit and it comes into the inner Solar System. The asteroid belt has gaps, known as Kirkwood gaps, where these resonances occur as the asteroids in these resonances have been moved onto other orbits.

New asteroids migrate into these resonances, due to the Yarkovsky effect that provides a continuing supply of near-Earth asteroids. Compared to the entire mass of the asteroid belt, the mass loss necessary to sustain the NEA population is relatively small; totaling less than 6% over the past 3.5 billion years. The composition of near-Earth asteroids is comparable to that of asteroids from the asteroid belt, reflecting a variety of asteroid spectral types.

A small number of NEAs are extinct comets that have lost their volatile surface materials, although having a faint or intermittent comet-like tail does not necessarily result in a classification as a near-Earth comet, making the boundaries somewhat fuzzy. The rest of the near-Earth asteroids are driven out of the asteroid belt by gravitational interactions with Jupiter.

Many asteroids have natural satellites (minor-planet moons). As of February 2019, 74 NEAs were known to have at least one moon, including three known to have two moons. The asteroid 3122 Florence, one of the largest PHAs with a diameter of 4.5 km (2.8 mi), has two moons measuring 100–300 m (330–980 ft) across, which were discovered by radar imaging during the asteroid's 2017 approach to Earth.

4.4 Some Well Known Asteroids

The best known and largest Asteroid is Ceres which would make a good location for a base in the Asteroid Belt

Ceres

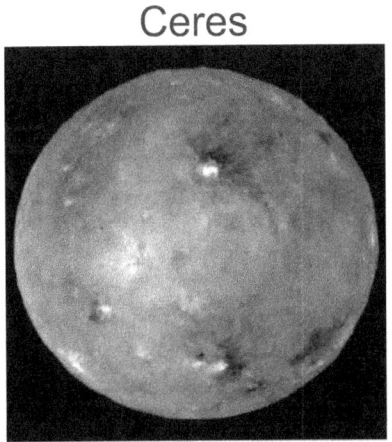

Ceres would be a great place to build a trading center and base in the Asteroid Belt. Many science fiction stories have underground cities on Ceres as a location for trade and relief for miners in the Belt.

A concept for a Colony on Ceres

Ceres is the smallest recognized dwarf planet, the closest dwarf planet to the Sun, and the largest object in the main asteroid belt that lies between the orbits of Mars and Jupiter. With a diameter of 940 km (580 mi), Ceres is both the largest of the asteroids and the only recognized dwarf planet inside Neptune's orbit. It's the 25th-largest body in the Solar System within the orbit of Neptune.

Ceres was the first asteroid to be discovered (by Giuseppe Piazzi at Palermo Astronomical Observatory on 1 January 1801). It was originally considered a planet, but was reclassified as an asteroid in the 1850s after many other objects in similar orbits were discovered.

Ceres is the only object in the asteroid belt rounded by its own gravity, although Vesta and perhaps other asteroids were so in the past. From Earth, the apparent magnitude of Ceres ranges from 6.7 to 9.3, peaking once at opposition every 15 to 16 months, which is its synodic period. Thus even at its brightest, it is too dim to be seen by the naked eye, except under extremely dark skies. Ceres has been classified as a C-type asteroid and, due to the presence of clay minerals, as a G-type asteroid.

Ceres appears to be partially differentiated into a muddy (ice-rock) mantle/core and a less-dense but stronger crust that is at most 30 percent ice. It probably no longer has an internal ocean of liquid water, but there is brine that can flow through the outer mantle and reach the surface. The surface is a mixture of water ice and various hydrated minerals such as carbonates and clay. Cryovolcanoes such as Ahuna Mons form at the rate of about one every fifty million years. In January 2014, emissions of water vapor were detected from several regions of Ceres. This was unexpected because large bodies in the asteroid belt typically do not emit vapor, a hallmark of comets. The atmosphere, however, is transient and of the minimal kind known as an exosphere.

The robotic NASA spacecraft Dawn entered orbit around Ceres on 6 March 2015.

The picture is a close up of Ceres big bright spot which is thought to be made of ice or a type of salt.

Vesta

Vesta is one of the largest objects in the asteroid belt, with a mean diameter of 525 kilometers (326 mi). It was discovered by the German astronomer Heinrich Wilhelm Matthias Olbers on 29 March 1807 and is named after Vesta, the virgin goddess of home and hearth from Roman mythology.

Vesta is the second-largest asteroid, both by mass and by volume, after the dwarf planet Ceres. It constitutes an estimated 9% of the mass of the asteroid belt. It is only slightly larger than Pallas (about 5% by volume), but is 25% to 30% more massive. Vesta is the only known remaining rocky protoplanet (with a differentiated interior) of the kind that formed the terrestrial planets. Numerous fragments of Vesta were ejected by collisions one and two billion years ago that left two enormous craters occupying much of Vesta's southern hemisphere. Debris from these events has fallen to Earth as howardite–eucrite–diogenite (HED) meteorites, which have been a rich source of information about Vesta.

Pallas

Pallas is the second asteroid to have been discovered, after Ceres. Like Ceres, it is believed to have a mineral composition similar to carbonaceous chondrite meteorites, though significantly less hydrated than Ceres. It is the third-largest asteroid in the Solar System by both volume and mass, and is a likely remnant protoplanet. It is 79% the mass of Vesta and 22% the mass of Ceres, constituting an estimated 7% of the mass of the asteroid belt. Its volume is equivalent to a sphere 505 to 520 kilometers (314 to 323 mi) in diameter, 90–96% the volume of Vesta.

During the planetary formation era of the Solar System, objects grew in size through an accretion process to approximately the size of Pallas. Most of these 'protoplanets' were incorporated into the growth of larger bodies, which became the planets, whereas others were ejected by the planets or destroyed in collisions with each other. Pallas, Vesta and Ceres appear to be the only intact bodies from this early stage of planetary formation to survive within the orbit of Neptune.

When Pallas was discovered by the German astronomer Heinrich Wilhelm Matthäus Olbers on 28 March 1802, it was counted as a planet, as were other asteroids in the early 19th century. The discovery of many more asteroids after 1845 eventually led to the separate listing of 'minor' planets from 'major' planets, and the realization in the 1950s that such small bodies did not form in the same way as (other) planets led to the gradual abandonment of the term 'minor planet' in favor of 'asteroid' (or, for larger bodies such as Pallas, 'planetoid')

Exploring and Settling Our Huge Solar System

4.5 A Story of Building an Asteroid Base

From my Science Fiction Book "Visiting Many Universes" is a description of building an Asteroid base for living and as the base to expand for building an interstellar ship construction:

Deciding how to expand next was also in the works. His plan was to buy his own asteroid and move everything he had into it. Finding asteroids is easy, but looking for just the right one takes time. It was best to check the public database maintained on all asteroid finds. This database listed thousands of them. He would need to find the right one then pay a fee to the finder to take possession. Spending a few weeks on the project he narrowed down the choice to a rock which was a few miles in diameter and five miles long. It was several million miles away around the belt from Ceres. He wanted lots of privacy for his next plans. Jason paid the rights holder some hundreds of thousands of credits. The sellers were happy—he paid a good amount—but not something outrageous which would be noticeable.

Then he used his ship to make a short visit out there to review the rock and plan how to use it. His idea was to hollow out some tunnels, have a landing bay, and a good size interior volume which would take time to complete. He would live in some rooms off the tunnel as the large open area was dugout.

When Jason got back to Ceres he hired a mining company which also did special projects to start cutting into the rock and building the basic structure. It would take them some months to make it livable. In the meantime he would start building his empire. It took six months to dig enough tunnels to make the place livable although it would take a couple of years to hollow out his entire design. The work was mainly done with robots which were guided on their routes by AI software, but it still took a long time to dig the large cavern.

He wanted a main cavern which could become an Earth like retreat for him and which was a mile wide, a mile long, and a ceiling which was over a thousand feet high. This was a lot of rock to remove. But it could be done with a team of robots which used lasers to break up the rock and different ore carrying robots which made a continuous stream of rock chips being dumped out of an industrial airlock at the end of the exit tunnel.

At the end of six months He moved into his partially finished new home. He had a bedroom, living room, kitchen, and miscellaneous room which had exercise equipment, his office, and a small lab. This would have to do until the main cavern was finished.

Aurora and Jason had calls daily. They were using an advanced model of the superluminal communicators which had a large bandwidth and allowed instant video communications from the asteroid home to Earth.

They were working on the research project into the gravity nullifier. If they could crack this technology then they would be able to create another multi-billion dollar industry and basically change the way human civilized technology worked. They had staffed a large laboratory studying gravity by this time with lots of world class experts working there. There was already a lot of Earth based research on gravity and they were hopeful that they could incorporate the alien gravity nullifier into existing technology development.

By using electron microscopes they were again able to deduce what components made up the nullifier. This included a strangely shaped component which turned out to be made of many metals with grains each oriented in strange patterns. Their scientists finally worked out that this core component blended electromagnetic energy with other "Flavors" of energy to create different gravity levels in the surrounding volumes. These gravity levels could be set from zero all the way to one hundred gravities depending on the power available. (It might go higher but they had only been able to test to that level of intensity).

This device when commercialized would create a revolution in transportation. Imagine take large objects from the surface of the Earth into space with one thousandth of the energy expended to do so today with rockets. It would also allow humanity to explore high gravity environments like Jupiter where the gravity was two and a half times Earth Gravity or even much higher gravity levels. They could also provide artificial gravity on space ships instead of using centrifugal force to anchor people to the surface. Weightlessness caused a lot of bone density problems and being able to alleviate that would save billions in medical care and specialized exercise equipment.

The bottom line was that in the next six months they would be able to start introducing and selling some null gravity devices. Then Jason would become the wealthiest man in the solar system.

4.6 An Asteroid Mining Base

This is drawing and concept of an asteroid mining colony attached to an asteroid to be mined.

Exploring and Settling Our Huge Solar System

5.0 The Kuiper Belt

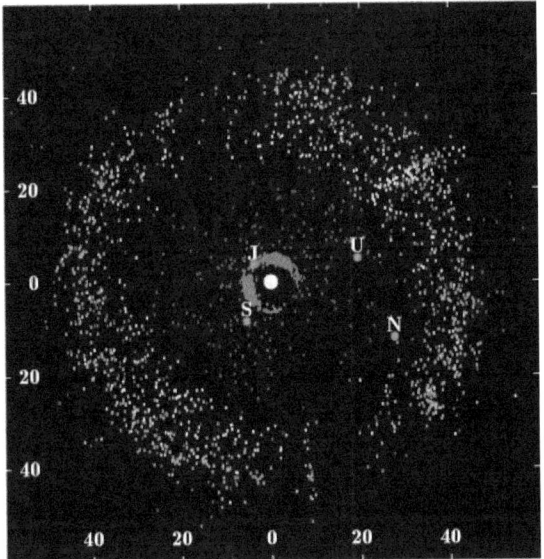

The Kuiper belt is a circumstellar disc in the outer Solar System, extending from the orbit of Neptune at 30 astronomical units (AU) to approximately 50 AU from the Sun. It is similar to the asteroid belt, but is far larger – 20 times as wide and 20–200 times as massive. Like the asteroid belt, it consists mainly of small bodies or remnants from when the Solar System formed. While many asteroids are composed primarily of rock and metal, most Kuiper belt objects are composed largely of frozen volatiles (termed "ices"), such as methane, ammonia and water. The Kuiper belt is home to three officially recognized dwarf planets: Pluto, Haumea and Makemake. Some of the Solar System's moons, such as Neptune's Triton and Saturn's Phoebe, may have originated in the region.

The Kuiper belt was named after Dutch-American astronomer Gerard Kuiper, though he did not predict its existence. In 1992, minor planet Albion was discovered, the first Kuiper belt object (KBO) since Pluto and Charon.

Since its discovery, the number of known KBOs has increased to thousands, and more than 100,000 KBOs over 100 km (62 mi) in diameter are thought to exist. The Kuiper belt was initially thought to be the main repository for periodic comets, those with orbits lasting less than 200 years. Studies since the mid-1990s have shown that the belt is dynamically stable and that comets' true place of origin is the scattered disc, a dynamically active zone created by the outward motion of Neptune 4.5 billion years ago; scattered disc objects such as Eris have extremely eccentric orbits that take them as far as 100 AU from the Sun.

The Kuiper belt is distinct from the theoretical Oort cloud, which is a thousand times more distant and is mostly spherical. The objects within the Kuiper belt, together with the members of the scattered disc and any potential Hills cloud or Oort cloud objects, are collectively referred to as trans-Neptunian objects (TNOs). Pluto is the largest and most massive member of the Kuiper belt, and the largest and the second-most-massive known TNO, surpassed only by Eris in the scattered disc. Originally considered a planet, Pluto's status as part of the Kuiper belt caused it to be reclassified as a dwarf planet in 2006. It is compositionally similar to many other objects of the Kuiper belt and its orbital period is characteristic of a class of KBOs, known as "plutinos",that share the same 2:3 resonance with Neptune.

The Kuiper belt and Neptune may be treated as a marker of the extent of the Solar System, alternatives being the heliopause and the distance at which the Sun's gravitational influence is matched by that of other stars (estimated to be between 50000 AU and about 2 light-years).

5.1 Some Kuiper Belt Objects

The Kuiper Belt has many thousands of asteroid like objects. Here are just a few which are fairly well known:

Ultima Thule

Ultima Thule is a trans-Neptunian object located in the Kuiper belt. It is a contact binary 36 km (22 mi) long, composed of two planetesimals 21 km (13 mi) and 15 km (9 mi) across, that are joined along their major axes. The larger lobe, which is flatter than smaller lobe, appears to be an aggregate of 8 or so smaller units, each approximately 5 km (3 mi) across that fused together before they came into contact. Because there have been few to no disruptive impacts on Arrokoth since it formed, the details of its formation have been preserved. With the New Horizons space probe's flyby at 05:33 on 1 January 2019 (UTC time), Arrokoth became the farthest and most primitive object in the Solar System visited by a spacecraft. At the time of the New Horizons flyby, the object had been nicknamed Ultima Thule.

Arrokoth was discovered on 26 June 2014 by astronomer Marc Buie and the New Horizons Search Team using the Hubble Space Telescope as part of a search for a Kuiper belt object for the New Horizons mission to target in its first extended mission; it was chosen over two other candidates to become the primary target of the mission. With an orbital period of about 298 years and a low orbital inclination and eccentricity, Arrokoth is classified as a cold classical Kuiper belt object.

Sedna

90377 Sedna, or simply Sedna, is a large planetoid in the outer reaches of the Solar System that was, as of 2020, at a distance of about 85 astronomical units (1.27×10^{10} km; 7.9×10^9 mi) from the Sun, about three times farther than Neptune. Spectroscopy has revealed that Sedna's surface composition is similar to those of some other trans-Neptunian objects, being largely a mixture of water, methane, and nitrogen ices with tholins. Its surface is one of the reddest among Solar System objects. It is a possible dwarf planet. Sedna is approximately tied with 2002 MS4 and 2002 AW197 as the largest planetoid not known to have a moon.

For most of its orbit, it is even farther from the Sun than at present, with its aphelion estimated at 937 AU (31 times Neptune's distance, or about 1.5% of a light-year), making it one of the most distant-known objects in the Solar System other than long-period comets.

Sedna has an exceptionally long and elongated orbit, taking approximately 11,400 years to complete and a distant point of closest approach to the Sun at 76 AU. These facts have led to much speculation about its origin.

The Minor Planet Center currently places Sedna in the scattered disc, a group of objects sent into highly elongated orbits by the gravitational influence of Neptune.

This classification has been contested because its perihelion is too large for it to have been scattered by a known planet, leading some astronomers to informally refer to it as the first known member of the inner Oort cloud.

Others speculate that it might have been tugged into its current orbit by a passing star, perhaps one within the Sun's birth cluster (an open cluster), or even that it was captured from another star system. Another hypothesis suggests that its orbit may be evidence for a large planet beyond the orbit of Neptune.

Haumea

Haumea (minor-planet designation 136108 Haumea) is a dwarf planet located beyond Neptune's orbit. It was discovered in 2004 by a team headed by Mike Brown of Caltech at the Palomar Observatory in the United States and independently in 2005 by a team headed by José Luis Ortiz Moreno at the Sierra Nevada Observatory in Spain, though the latter claim has been contested. On September 17, 2008, it was named after Haumea, the Hawaiian goddess of childbirth, under the expectation by the International Astronomical Union (IAU) that it would prove to be a dwarf planet. It is probably the third-largest known trans-Neptunian object, after Eris and Pluto.

Haumea's mass is about one-third that of Pluto, and 1/1400 that of Earth. Although its shape has not been directly observed, calculations from its light curve are consistent with it being a Jacobi ellipsoid (the shape it would be if it were a dwarf planet), with its major axis twice as long as its minor. In October 2017, astronomers announced the discovery of a ring system around Haumea, representing the first ring system discovered for a trans-Neptunian object. Haumea's gravity was until recently thought to be sufficient for it to have relaxed into hydrostatic equilibrium, though that is now unclear. Haumea's elongated shape together with its rapid rotation, rings, and high albedo (from a surface of crystalline water ice), are thought to be the consequences of a giant collision, which left Haumea the largest member of a collisional family that includes several large trans-Neptunian objects and Haumea's two known moons,

Hi'iaka and Namaka.

Makemake

Makemake (minor-planet designation 136472 Makemake) is a likely dwarf planet and perhaps the second largest Kuiper belt object in the classical population, with a diameter approximately two-thirds that of Pluto. Makemake has one known satellite. It's extremely low average temperature, about 40 K (−230 °C), means its surface is covered with methane, ethane, and possibly nitrogen ices.

Makemake was discovered on March 31, 2005, by a team led by Michael E. Brown, and announced on July 29, 2005. Initially, it was known as 2005 FY9 and later given the minor-planet number 136472. In July 2008 it was named after Makemake, the creator god of the Rapa Nui people of Easter Island, under the expectation by the International Astronomical Union (IAU) that it would prove to be a dwarf planet.

Exploring and Settling Our Huge Solar System

5.2 Building Bases in the Kuiper Belt

First of all, why would we build anything so far out in the Solar System?

The main reason I can think of is as a centralized hub for mining activities in the Kuiper Belt. Since the belt is in a huge ring around the Solar System there might need to be multiple bases built around the Solar System in support of the dispersed mining activities.

Things to be mined could include water ice, metals like iron, and rare earths.

What a base and mining site on these objects might look like

6.0   The Oort Cloud

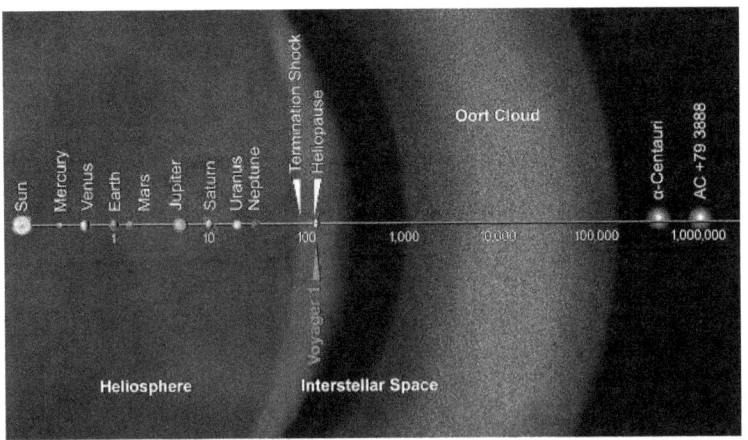

The Oort cloud was first described in 1950 by Dutch astronomer Jan Oort is a theoretical cloud of predominantly icy planetesimals proposed to surround the Sun at distances ranging from 2,000 to 200,000 au (0.03 to 3.2 light-years). It is divided into two regions: a disc-shaped inner Oort cloud (or Hills cloud) and a spherical outer Oort cloud. Both regions lie beyond the heliosphere and in interstellar space. The Kuiper belt and the scattered disc, the other two reservoirs of trans-Neptunian objects, are less than one thousandth as far from the Sun as the Oort cloud.

This cloud may have millions of objects in it due to its size.

The outer limit of the Oort cloud defines the cosmographic boundary of the Solar System and the extent of the Sun's Hill sphere. The outer Oort cloud is only loosely bound to the Solar System, and thus is easily affected by the gravitational pull both of passing stars and of the Milky

Way itself. These forces occasionally dislodge comets from their orbits within the cloud and send them toward the inner Solar System. Based on their orbits, most of the short-period comets may come from the scattered disc, but some may still have originated from the Oort cloud.

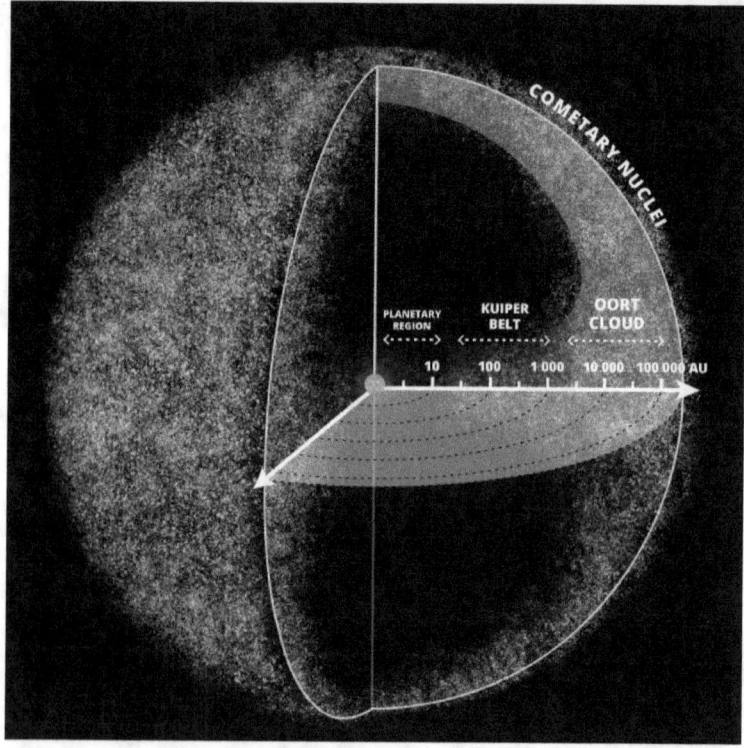

The shape of the OORT Cloud is thought to be spherical even though the Kuiper Belt is mainly in the flat plane of the rest of the Solar System.

7.0   Comets Around the Sun

Comets are all over the Solar System and most are thought to come from the Oort Cloud. Could a comet be mined or settled? We don't know enough to answer this yet.

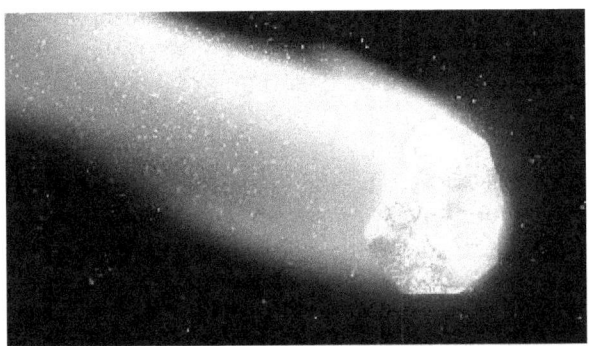

A comet is an icy, small Solar System body that, when passing close to the Sun, warms and begins to release gases, a process called outgassing. This produces a visible atmosphere or coma, and sometimes also a tail. These phenomena are due to the effects of solar radiation and the solar wind acting upon the nucleus of the comet.

Comet nuclei range from a few hundred meters to tens of kilometers across and are composed of loose collections of ice, dust, and small rocky particles. The coma may be up to 15 times Earth's diameter, while the tail may stretch beyond one astronomical unit. If sufficiently bright, a comet may be seen from Earth without the aid of a telescope and may subtend an arc of 30° (60 Moons) across the sky. Comets have been observed and recorded since ancient times by many cultures.

Comets usually have highly eccentric elliptical orbits, and they have a wide range of orbital periods, ranging from several years to potentially several millions of years. Short-period comets originate in the Kuiper belt or its associated scattered disc, which lie beyond the orbit of Neptune. Long-period comets are thought to originate in the Oort cloud, a spherical cloud of icy bodies extending from outside the Kuiper belt to halfway to the nearest star. Hyperbolic comets may pass once through the inner Solar System before being flung to interstellar space. The appearance of a comet is called an apparition.

Comets are distinguished from asteroids by the presence of an extended, gravitationally unbound atmosphere surrounding their central nucleus. This atmosphere has parts termed the coma (the central part immediately surrounding the nucleus) and the tail (a typically linear section consisting of dust or gas blown out from the coma by the Sun's light pressure or outstreaming solar wind plasma). However, extinct comets that have passed close to the Sun many times have lost nearly all of their volatile ices and dust and may come to resemble small asteroids. Asteroids are thought to have a different origin from comets, having formed inside the orbit of Jupiter rather than in the outer Solar System. The discovery of main-belt comets and active centaur minor planets has blurred the distinction between asteroids and comets. In the early 21st century, the discovery of some minor bodies with long-period comet orbits, but characteristics of inner solar system asteroids, were called Manx comets. They are still classified as comets, such as C/2014 S3

As of July 2019 there are 6,619 known comets, a number that is steadily increasing as more are discovered. However, this represents only a tiny fraction of the total potential comet population, as the reservoir of comet-like bodies in the outer Solar System (in the Oort cloud) is estimated to be one trillion. Roughly one comet per year is visible to the naked eye, though many of those are faint and unspectacular. Particularly bright examples are called "great comets". Comets have been visited by unmanned probes such as the European Space Agency's Rosetta, which became the first to land a robotic spacecraft on a comet, and NASA's Deep Impact, which blasted a crater on Comet Tempel 1 to study its interior.

7.1   Halley's Comet

Halley's Comet is the most famous of all the comets and has been observed by humanity going back thousands of years. It is 15 km X 8km so is not that big. But it might be a great place to build a base with a long periodicity of travel on a large parabola throughout the Solar System

Halley's Comet or Comet Halley, officially designated 1P/Halley, is a short-period comet visible from Earth every 75–76 years. Halley is the only known short-period comet that is regularly visible to the naked eye from Earth, and the only naked-eye comet that can appear twice in a human lifetime. Halley last appeared in the inner parts of the Solar System in 1986 and will next appear in mid-2061.

Halley's periodic returns to the inner Solar System have been observed and recorded by astronomers around the world since at least 240 BC. But it was not until 1705 that the English astronomer Edmond Halley understood that these appearances were reappearances of the same comet. As a result of this discovery, the comet is now named after Halley.

During its 1986 apparition, Halley's Comet became the first comet to be observed in detail by spacecraft, providing the first observational data on the structure of a comet nucleus and the mechanism of coma and tail formation. These observations supported a number of longstanding hypotheses about comet construction, particularly Fred Whipple's "dirty snowball" model, which correctly predicted that Halley would be composed of a mixture of volatile ices—such as water, carbon dioxide, and ammonia—and dust. The missions also provided data that substantially reformed and reconfigured these ideas; for instance, it is now understood that the surface of Halley is largely composed of dusty, non-volatile materials, and that only a small portion of it is icy.

8.0   Summary

After reading this book I hope you understand why the more I learned the more awed I became about the huge size and potential of the Solar System.

The human race can settle and use the resources of the Solar System to advance human civilization for hundreds and even thousands of years.

I'm also a big proponent of Space Colonies and have even written the book "Designing and Building Space Colonies". With robotics and three dimensional print manufacturing we could also use the resources of the Solar System to build many huge space colonies orbiting different planets and moons.

There are literally millions of objects the size of asteroids which can be settled or mined for their resources. There is much more in the way of resources than we ever thought. As we learn about the true resources of our system we become more and more awed of the results of our exploration missions.

Overall, we should help space entrepreneurs to explore and settle our Solar System since the eventual results will be huge.

Sincerely,

Martin K. Ettington

March 2021

8.0 Index

9.0 Appendix-All Solar System Moons & Selected Objects

The Earth and its Moon

Body	Description	Date of Discovery	Discovery Location	Discoverer
Earth	The name Earth comes from the Indo-European base 'er,'which produced the Germanic noun 'ertho,' and ultimately German 'erde,' Dutch 'aarde,' Scandinavian 'jord,' and English 'earth.' Related forms include Greek 'eraze,' meaning 'on the ground,' and Welsh 'erw,' meaning 'a piece of land.'	?	?	?
Earth I (Moon)	Every civilization has had a name for the satellite of Earth that is known, in English, as the Moon. The Moon is known as Luna in Italian, Latin, and Spanish, as Lune in French, as Mond in German, and as Selene in Greek.	?	?	?

Mars and its Moons

The names of the moons of Mars and the English translations of the names were specifically proposed by their discoverer, Asaph Hall, and as such, they have been accepted and retained under the current IAU nomenclature.

Body	Description	Date of Discovery	Discovery Location	Discoverer
Mars	Named by the Romans for their god of war because of its red, bloodlike color. Other civilizations also named this planet from this attribute; for example, the Egyptians named it "Her Desher," meaning "the red one."	?	?	?
Mars I (Phobos)	Inner satellite of Mars. Named for one of the horses that drew Mars' chariot; also called an "attendant" or "son" of Mars, according to	August 17, 1877	Washington	A. Hall

Body	Description	Date of Discovery	Discovery Location	Discoverer
	chapter 15, line 119 of Homer's "Iliad." This Greek word means "flight."			
Mars II (Deimos)	This outer Martian satellite was named for one of the horses that drew Mars' chariot; also called an "attendant" or "son" of Mars, according to chapter 15, line 119 of Homer's "Iliad." Deimos means "fear" in Greek.	August 11, 1877	Washington	A. Hall

Selected Asteroids (of the Main Belt) and their Satellites

Body	Description	Date of Discovery	Discovery Location	Discoverer
(433) Eros	Named for the Greek god of love.	August 13, 1898	Berlin	C.G. Witt
(951) Gaspra	Named for a resort on the Crimean Peninsula.	July 30, 1916	Simeis	G. Neujmin
(243) Ida	Named for a nymph who raised the infant Zeus. Ida is also the name of a mountain on the island of Crete, the location of the cave where Zeus was reared.	September 29, 1884	Vienna	J. Palisa
(243) Ida I (Dactyl)	Named for a group of mythological beings who lived on Mount Ida, where the infant Zeus was hidden and raised (according to some accounts) by the nymph Ida.	August 28, 1993		Galileo imaging and infrared science teams.
(253) Mathilde	The name was suggested by a staff member of the Paris Observatory who first computed an orbit for	November 12, 1885	Vienna	J. Palisa

Body	Description	Date of Discovery	Discovery Location	Discoverer
	Mathilde. The name is thought to honor the wife of the vice director of the Paris Observatory at that time.			
(22) Kalliope I (Linus)	Satellite of (22) Kalliope. In various accounts of Greek mythology, Linus is considered to be the son of the Muse Kalliope and the inventor of melody and rhythm.	August 29 and September 2, 2001	Mauna Kea	J.-L. Margot, M.E. Brown, W.J. Merline, F. Menard, L. Close, C. Dumas, C.R. Chapman, and D.C. Slater
(45) Eugenia I (Petit-Prince)	Satellite of (45) Eugenia. The Little Prince, Napolean-Eugene-Louis-Jean-Joseph Bonaparte (1856-1879), was the son of Eugenia de Montijo de Guzm\'an and Napoleon III.	November 1, 1998	Mauna Kea	W.J. Merline, L. Close, C. Dumas, C.R. Chapman, F. Roddier, F. Menard, D.C. Slater, G. Duvert, C. Shelton, and T. Morgan

Jupiter and its Moons

Satellites in the Jovian system are named for Zeus/Jupiter's lovers and descendants. Names of outer satellites with a prograde orbit generally end with the letter "a" (although an "o" ending has been reserved for some unusual cases), and names of satellites with a retrograde orbit end with an "e."

Body	Description	Date of Discovery	Discovery Location	Discoverer
Jupiter	The largest and most massive of the planets was named Zeus by the Greeks and Jupiter by the Romans; he was the most important deity in both pantheons.	?	?	?
Jupiter I (Io)	Io, the daughter of Inachus, was changed by Jupiter into a cow to protect	January 8, 1610	Padua	Galileo (Simon Marius probably made an independent

	her from Hera's jealous wrath. But Hera recognized Io and sent a gadfly to torment her. Io, maddened by the fly, wandered throughout the Mediterranean region.			discovery of the Galilean satellites at about the same time that Galileo did, and he may have unwittingly sighted them up to a month earlier, but the priority must go to Galileo because he published his discovery first.)
Jupiter II (Europa)	Beautiful daughter of Agenor, king of Tyre, she was seduced by Jupiter, who had assumed the shape of a white bull. When Europa climbed on his back he swam with her to Crete, where she bore several children, including Minos.	January 8, 1610	Padua	Galileo (who evidently observed the combined image of Io and Europa the previous night)
Jupiter III (Ganymede)	Beautiful young boy who was carried to Olympus by Jupiter disguised as an eagle. Ganymede then became the cupbearer of the Olympian gods.	January 7, 1610	Padua	Galileo
Jupiter IV (Callisto)	Beautiful daughter of Lycaon, she was seduced by Jupiter, who changed her into a bear to protect her from Hera's jealousy.	January 7, 1610	Padua	Galileo
Jupiter V (Amalthea)	A naiad who nursed the new-born Jupiter.	September 9, 1892	Mt. Hamilton	E.E. Barnard

	She had as a favorite animal a goat which is said by some authors to have nourished Jupiter. The name was suggested by Flammarion.			
Jupiter VI (Himalia)	A Rhodian nymph who bore three sons of Zeus.	December 4, 1904	Mt. Hamilton	C.D. Perrine
Jupiter VII (Elara)	Daughter of King Orchomenus, a paramour of Zeus, and by him the mother of the giant Tityus.	January 3, 1905	Mt. Hamilton	C.D. Perrine
Jupiter VIII (Pasiphae)	Wife of Minos, king of Crete. Zeus made approaches to her as a bull (taurus). She then gave birth to the Minotaur. (Spelling changed from Pasiphaë to Pasiphae July 2009.)	January 27, 1908	Greenwich	P.J. Melotte
Jupiter IX (Sinope)	Daughter of the river god Asopus. Zeus desired to make love to her. Instead of this he granted perpetual virginity, after he had been deceived by his own promises. (In the same way, she also fooled Apollo.)	July 21, 1914	Mt. Hamilton	S.B. Nicholson
Jupiter X (Lysithea)	Daughter of Kadmos, also named Semele, mother of Dionysos by Zeus.	July 6, 1938	Mt. Wilson	S.B. Nicholson

	According to others, she was the daughter of Evenus and mother of Helenus by Jupiter.			
Jupiter XI (Carme)	A nymph and attendant of Artemis; mother, by Zeus, of Britomartis.	July 30, 1938	Mt. Wilson	S.B. Nicholson
Jupiter XII (Ananke)	Goddess of fate and necessity, mother of Adrastea by Zeus.	September 28, 1951	Mt. Wilson	S.B. Nicholson
Jupiter XIII (Leda)	Seduced by Zeus in the form of a swan, she was the mother of Pollux and Helen.	September 11, 1974	Palomar	C.T. Kowal
Jupiter XIV (Thebe)	An Egyptian king's daughter, granddaughter of Io, mother of Aigyptos by Zeus. The Egyptian city of Thebes was named after her.	March 5, 1979	Voyager 1	Voyager Science Team
Jupiter XV (Adrastea)	A nymph of Crete to whose care Rhea entrusted the infant Zeus.	July, 1979	Voyager 2	Voyager Science Team
Jupiter XVI (Metis)	First wife of Zeus. He swallowed her when she became pregnant; Athena was subsequently born from the forehead of Zeus.	March 4, 1979	Voyager 1	Voyager Science Team

Jupiter XVII (Callirrhoe)	Daughter of the river god Achelous and stepdaughter of Zeus.	October 19, 1999	Spacewatch	J.V. Scotti, T.B. Spahr, R.S. McMillan, J.A. Larson, J. Montani, A.E. Gleason, and T. Gehrels
Jupiter XVIII (Themisto)	Daughter of the Arcadian river god Inachus, mother of Ister by Zeus.	September 30, 1975, rediscovered November 21, 2000	Palomar, rediscovered at Mauna Kea	C.T. Kowal and E. Roemer (1975), and S.S. Sheppard, D.C. Jewitt, Y.R. Fernandez, G. Magnier, M. Holman, B.G. Marsden, and G.V. Williams (2000).
Jupiter XIX (Megaclite)	Daughter of Macareus, who with Zeus gave birth to Thebe and Locrus.	November 25, 2000	Mauna Kea	S.S. Sheppard, D.C. Jewitt, Y.R. Fernandez, and G. Magnier
Jupiter XX (Taygete)	Daughter of Atlas, one of the Pleiades, mother of Lakedaimon by Zeus.	November 25, 2000	Mauna Kea	S.S. Sheppard, D.C. Jewitt, Y.R. Fernandez, and G. Magnier
Jupiter XXI (Chaldene)	Bore the son Solymos with Zeus.	November 26, 2000	Mauna Kea	S.S. Sheppard, D.C. Jewitt, Y.R. Fernandez, and G. Magnier
Jupiter XXII (Harpalyke)	Daughter and wife of Clymenus. In revenge for this incestuous relationship, she killed the son she bore him, cooked the corpse, and served it to Clymenus. She was transformed into	November 23, 2000	Mauna Kea	S.S. Sheppard, D.C. Jewitt, Y.R. Fernandez, and G. Magnier

	the night bird called Chalkis, and Clymenus hanged himself. Some say that she was transformed into that bird because she had intercourse with Zeus.			
Jupiter XXIII (Kalyke)	Nymph who bore the handsome son Endymion with Zeus.	November 23, 2000	Mauna Kea	S.S. Sheppard, D.C. Jewitt, Y.R. Fernandez, and G. Magnier
Jupiter XXIV (Iocaste)	Wife of Laius, King of Thebes, and mother of Oedipus. After Laius was killed, Iocaste unknowingly married her own son Oedipus. When she learned that her husband was her son, she killed herself. Some say she was the mother of Agamedes by Zeus.	November 23, 2000	Mauna Kea	S.S. Sheppard, D.C. Jewitt, Y.R. Fernandez, and G. Magnier
Jupiter XXV (Erinome)	Daughter of Celes, compelled by Venus to fall in love with Jupiter.	November 23, 2000	Mauna Kea	S.S. Sheppard, D.C. Jewitt, Y.R. Fernandez, and G. Magnier
Jupiter XXVI (Isonoe)	A Danaid, bore with Zeus the son Orchomenos.	November 23, 2000	Mauna Kea	S.S. Sheppard, D.C. Jewitt, Y.R. Fernandez, and G. Magnier
Jupiter XXVII (Praxidike)	Goddess of punishment, mother of Klesios by Zeus.	November 23, 2000	Mauna Kea	S.S. Sheppard, D.C. Jewitt, Y.R. Fernandez, and G. Magnier

Jupiter XXVIII (Autonoe)	Mother of the Graces by Zeus according to some authors.	December 10, 2001	Mauna Kea	S.S. Sheppard, D.C. Jewitt and J. Kleyna
Jupiter XXIX (Thyone)	Semele, mother of Dionysos by Zeus. She received the name of Thyone in Hades by Dionysos before he ascended up with her from there to heaven.	December 11, 2001	Mauna Kea	S.S. Sheppard, D.C. Jewitt and J. Kleyna
Jupiter XXX (Hermippe)	Consort of Zeus and mother of Orchomenos by him.	December 9, 2001	Mauna Kea	S.S. Sheppard, D.C. Jewitt and J. Kleyna
Jupiter XXXI (Aitne)	A Sicilian nymph, conquest of Zeus.	December 9, 2001	Mauna Kea	S.S. Sheppard, D.C. Jewitt and J. Kleyna
Jupiter XXXII (Eurydome)	Mother of the Graces by Zeus, according to some authors. (Source: Cornutus: Theologiae Graecae compendium 15)	December 9, 2001	Mauna Kea	S.S. Sheppard, D.C. Jewitt and J. Kleyna
Jupiter XXXIII (Euanthe)	The mother of the Graces by Zeus, according to some authors.	December 11, 2001	Mauna Kea	S.S. Sheppard, D.C. Jewitt and J. Kleyna
Jupiter XXXIV (Euporie)	One of the Horae, a daughter of Zeus and Themis.	December 11, 2001	Mauna Kea	S.S. Sheppard, D.C. Jewitt and J. Kleyna
Jupiter XXXV (Orthosie)	One of the Horae, a daughter of Zeus and Themis.	December 11, 2001	Mauna Kea	S.S. Sheppard, D.C. Jewitt and J. Kleyna

Jupiter XXXVI (Sponde)	One of the Horae (Seasons), daughter of Zeus.	December 9, 2001	Mauna Kea	S.S. Sheppard, D.C. Jewitt and J. Kleyna
Jupiter XXXVII (Kale)	One of the Graces, a daughter of Zeus, husband of Hephaistos.	December 9, 2001	Mauna Kea	S.S. Sheppard, D.C. Jewitt and J. Kleyna
Jupiter XXXVIII (Pasithee)	One of the Graces, a daughter of Zeus.	December 11, 2001	Mauna Kea	S.S. Sheppard, D.C. Jewitt and J. Kleyna
Jupiter XXXIX (Hegemone)	One of the Graces, a daughter of Zeus.	February 8, 2003	Mauna Kea	S.S. Sheppard
Jupiter XL (Mneme)	One of the Muses, a daughter of Zeus.	February 9, 2003	Mauna Kea	B. Gladman and L. Allen
Jupiter XLI (Aoede)	One of the Muses, a daughter of Zeus.	February 8, 2003	Mauna Kea	S.S. Sheppard
Jupiter XLII (Thelxinoe)	One of the Muses, a daughter of Zeus.	February 9, 2003	Mauna Kea	S.S. Sheppard
Jupiter XLIII (Arche)	One of the Muses, a daughter of Zeus.	October 31, 2002	Mauna Kea	S.S. Sheppard
Jupiter XLIV (Kallichore)	One of the Muses, a daughter of Zeus.	February 6, 2003	Mauna Kea	S.S. Sheppard
Jupiter XLV (Helike)	One of the Muses, a daughter of Zeus.	February 6, 2003	Mauna Kea	S.S. Sheppard
Jupiter XLVI (Carpo)	One of the Horae, a daughter of Zeus.	February 26, 2003	Mauna Kea	S.S. Sheppard

Jupiter XLVII (Eukelade)	One of the Muses, a daughter of Zeus.	February 5, 2003	Mauna Kea	S.S. Sheppard
Jupiter XLVIII (Cyllene)	Daughter of Zeus, a nymph.	February 9, 2003	Mauna Kea	S.S. Sheppard
Jupiter XLIX (Kore)	Daughter of Zeus and Demeter, also known as Persephone.	February 8, 2003	Mauna Kea	S. Sheppard, D.C. Jewitt, J. Kleyna
Jupiter L (Herse)	Daughter of Zeus and divine moon (Selene).	February 27, 2003	Mauna Kea	B. Gladman, J. Kavelaars, J.-M. Petit, and L. Allen
Jupiter LI (unnamed)		September 7, 2010	Palomar	R. Jacobson, M. Brozovic, B. Gladman, M. Alexandersen
Jupiter LII (unnamed)		September 8, 2010	Mauna Kea	C. Veillet
Jupiter LIII (Dia)	Greek meaning "She who belongs to Zeus". Dia is the daughter of Eioneus known as the divine daughter of the seashore. Zeus, disguised as a stallion, seduced Dia, who then gave birth to Peirithous."	December 5, 2000	Mauna Kea	S. S. Sheppard, D. C. Jewitt, Y. R. Fernandez, and G. Magnier
Jupiter LXII (Valetudo)	Great-granddaughter of Jupiter. Roman name for Greek Hygeia. She is the goddess of health	March 23, 2017	Cerro Tololo	S. S. Sheppard

Body	Description	Date of Discovery	Discovery Location	Discoverer
	and hygiene.			
Jupiter LXV (Pandia)	Daughter of Zeus and the Moon goddess Selene, goddess of the full moon, and sister of Ersa.	March 23, 2017	Cerro Tololo	S. S. Sheppard
Jupiter LXXI (Ersa)	Daughter of Zeus and the Moon goddess Selene, goddess of the dew, and sister of Pandia.	May 11, 2018	Cerro Tololo	S. S. Sheppard

Saturn and its Moons

Satellites in the saturnian system are named for Greco-Roman titans, descendants of the titans, the Roman god of the beginning, and giants from Greco-Roman and other mythologies. Gallic, Inuit and Norse names identify three different orbit inclination groups, where inclinations are measured with respect to the ecliptic, not Saturn's equator or orbit. Retrograde satellites (those with an inclination of 90 to 180 degrees) are named for Norse giants (except for Phoebe, which was discovered long ago and is the largest). Prograde satellites with an orbit inclination of around 36 degrees are named for Gallic giants, and prograde satellites with an orbit inclination of around 48 degrees are named for Inuit giants and spirits.

Note: 20 new moons discovered in 2019 and we are awaiting the official names to be selected before updating the table below.

Body	Description	Date of Discovery	Discovery Location	Discoverer
Saturn	Roman name for the Greek Cronos, father of Zeus/Jupiter. Other civilizations have given different names to Saturn, which is the farthest planet from Earth that can be observed by the naked human eye. Most of its satellites were named for Titans	?	?	?

	who, according to Greek mythology, were brothers and sisters of Saturn.			
Saturn I (Mimas)	Named by Herschel's son John in the early 19th century for a Giant felled by Hephaestus (or Ares) in the war between the Titans and Olympian gods.	July 18, 1789	Slough	W. Herschel
Saturn II (Enceladus)	Named by Herschel's son John for the Giant Enceladus. Enceladus was crushed by Athene in the battle between the Olympian gods and the Titans. Earth piled on top of him became the island of Sicily.	August 28, 1789	Slough	W. Herschel
Saturn III (Tethys)	Cassini wished to name Tethys and the other three satellites that he discovered (Dione, Rhea, and Iapetus) for Louis XIV. However, the names used today for these satellites were applied in the early 19th century by John Herschel, who named them for Titans and Titanesses, brothers and sisters of Saturn. Tethys was the wife of Oceanus and mother of all rivers and Oceanids.	March 21, 1684	Paris	G.D. Cassini
Saturn IV	Dione was the sister	March 21,	Paris	G.D. Cassini

(Dione)	of Cronos and mother (by Zeus) of Aphrodite.	1684		
Saturn V (Rhea)	A Titaness, mother of Zeus by Kronos.	December 23, 1672	Paris	G.D. Cassini
Saturn VI (Titan)	Named by Huygens, who first called it "Luna Saturni." In Greek Mythology, a Giant, and one of two generations of immortal giants (Titans) of incredible strength and stamina who were overthrown by a race of younger gods, the Olympians.	March 25, 1655	The Hague	C. Huygens
Saturn VII (Hyperion)	Named by Lassell for one of the Titans.	September 16, 1848	Cambridge, MA	W.C. Bond and G.P. Bond; independently discovered September 18, 1848 at Liverpool by W. Lassell
Saturn VIII (Iapetus)	Named by John Herschel for one of the Titans.	October 25, 1671	Paris	G.D. Cassini
Saturn IX (Phoebe)	Named by Pickering for one of the Titanesses.	August 16, 1898	Arequipa	W.H. Pickering
Saturn X (Janus)	First reported (though with an incorrect orbital period) and named by A. Dollfus from observations in Dec. 1966, this satellite was finally	December 15, 1966 (Dollfus), February 19, 1980 (Pascu)	Pic du Midi (Dollfus), Washington (Pascu)	A. Dollfus (1966), D. Pascu (1980)

	confirmed in 1980. It was proven to have a twin, Epimetheus, sharing the same orbit but never actually meeting. It is named for the Roman god of the beginning. The two-faced god could look forward and backward at the same time.			
Saturn XI (Epimetheus)	First suspected by J. Fountain and S. Larson as confusing the detection of Janus. They assigned the correct orbital period, and the satellite was finally confirmed in 1980. Named for the son of the Titan Iapetus. In contrast with his far-sighted brother Prometheus, he "subsequently realized" that he was in the wrong.	1977 (Fountain and Larson), February 26, 1980 (Cruikshank)	Tucson (Fountain and Larson), Mauna Kea (Cruikshank)	J. Fountain and S. Larson (1977), D. Cruikshank (1980)
Saturn XII (Helene)	A granddaughter of Kronos, for her beauty she triggered off the Trojan War.	March 1, 1980	Pic du Midi	P. Laques and J. Lecacheux
Saturn XIII (Telesto)	Daughter of the Titans Oceanus and Tethys.	April 8, 1980	Tucson	B.A. Smith, H. Reitsema, S.M. Larson, and J. Fountain
Saturn XIV (Calypso)	Daughter of the Titans Oceanus and Tethys and paramour of Odysseus.	March 13, 1980	Flagstaff	D. Pascu, P.K. Seidelmann, W. Baum, and D. Currie

Saturn XV (Atlas)	A Titan; he held the heavens on his shoulders.	October 1980	Voyager 1	Voyager Science Team
Saturn XVI (Prometheus)	Son of the Titan Iapetus, brother of Atlas and Epimetheus, he gave many gifts to humanity, including fire.	October 1980	Voyager 1	Voyager Science Team
Saturn XVII (Pandora)	Made of clay by Hephaestus at the request of Zeus. She married Epimetheus and opened the box that loosed a host of plagues upon humanity.	October 1980	Voyager 1	Voyager Science Team
Saturn XVIII (Pan)	Greek god of pastoralism, he was half goat and half human. Son of Hermes, brother of Daphnis, and a descendant of the Titans. Discovered orbiting in the Encke division in Saturn's A ring.	1990	Voyager 2	M.R. Showalter
Saturn XIX (Ymir)	Ymir is the primordial Norse giant and the progenitor of the race of frost giants.	August 7, 2000	La Silla	B. Gladman, J. Kavelaars, J.-M. Petit, H. Scholl, M. Holman, B.G. Marsden, P. Nicholson and J.A. Burns
Saturn XX (Paaliaq)	Named for an Inuit giant.	August 7, 2000	La Silla	B. Gladman, J. Kavelaars, J.-M. Petit, H. Scholl, M.

				Holman, B.G. Marsden, P. Nicholson and J.A. Burns
Saturn XXI (Tarvos)	Named for a Gallic giant.	September 23, 2000	Mauna Kea	B. Gladman, J. Kavelaars, J.-M. Petit, H. Scholl, M. Holman, B.G. Marsden, P. Nicholson and J.A. Burns
Saturn XXII (Ijiraq)	Named for an Inuit giant.	September 23, 2000	Mauna Kea	B. Gladman, J. Kavelaars, J.-M. Petit, H. Scholl, M. Holman, B.G. Marsden, P. Nicholson and J.A. Burns
Saturn XXIII (Suttungr)	Named for a Norse giant who kindled flames that destroyed the world.	September 23, 2000	Mauna Kea	B. Gladman, J. Kavelaars, J.-M. Petit, H. Scholl, M. Holman, B.G. Marsden, P. Nicholson and J.A. Burns
Saturn XXIV (Kiviuq)	Named for an Inuit giant.	August 7, 2000	La Silla	B. Gladman, J. Kavelaars, J.-M. Petit, H. Scholl, M. Holman, B.G. Marsden, P. Nicholson and J.A. Burns
Saturn XXV (Mundilfari)	Named for an Norse giant.	September 23, 2000	Mauna Kea	B. Gladman, J. Kavelaars, J.-M. Petit, H. Scholl, M.

				Holman, B.G. Marsden, P. Nicholson and J.A. Burns
Saturn XXVI (Albiorix)	Named for a Gallic giant who was considered to be the king of the world.	November 9, 2000	Mt. Hopkins	M. Holman
Saturn XXVII (Skathi)	Named for a Norse giantess.	September 23, 2000	Mauna Kea	B. Gladman, J. Kavelaars, J.-M. Petit, H. Scholl, M. Holman, B.G. Marsden, P. Nicholson and J.A. Burns
Saturn XXVIII (Erriapus)	Named for a Gallic giant.	September 23, 2000	Mauna Kea	B. Gladman, J. Kavelaars, J.-M. Petit, H. Scholl, M. Holman, B.G. Marsden, P. Nicholson and J.A. Burns
Saturn XXIX (Siarnaq)	Named for an Inuit giant.	September 23, 2000	Mauna Kea	B. Gladman, J. Kavelaars, J.-M. Petit, H. Scholl, M. Holman, B.G. Marsden, P. Nicholson and J.A. Burns
Saturn XXX (Thrymr)	Named for a Norse giant.	September 23, 2000	Mauna Kea	B. Gladman, J. Kavelaars, J.-M. Petit, H. Scholl, M. Holman, B.G. Marsden, P. Nicholson and J.A. Burns

Exploring and Settling Our Huge Solar System

Saturn XXXI (Narvi)	Named for a Norse giant.	February 5, 2003	Mauna Kea	S.S. Sheppard, D.C. Jewitt, and J. Kleyna
Saturn XXXII (Methone)	One of the Alkyonides, the seven beautiful daughters of the Giant Alkyoneos.	June 1, 2004		Cassini Imaging Science Team
Saturn XXXIII (Pallene)	One of the Alkyonides, the seven beautiful daughters of the Giant Alkyoneos.	June 1, 2004		Cassini Imaging Science Team
Saturn XXXIV (Polydeuces)	Twin brother of Castor, son of Zeus and Leda.	October 21, 2004		Cassini Imaging Science Team
Saturn XXXV (Daphnis)	Shepherd, pipes player, and pastoral poet in Greek mythology. Son of Hermes, brother of Pan, and decendant of the Titans. Discovered orbiting in the Keeler gap in Saturn's A ring.	May 1, 2005		Cassini Imaging Science Team
Saturn XXXVI (Aegir)	Norse ocean giant who represents the peaceful sea, a stiller of storms.	December 12, 2004	Mauna Kea	S. Sheppard, D.C. Jewitt, J. Kleyna
Saturn XXXVII (Bebhionn)	Beautiful Celtic giantess.	December 12, 2004	Mauna Kea	S. Sheppard, D.C. Jewitt, J. Kleyna
Saturn XXXVIII (Bergelmir)	Norse frost giant, son of Ymir and one of the Hrimthursar, one of only two members of the frost giant race to escape being drowned	December 12, 2004	Mauna Kea	S. Sheppard, D.C. Jewitt, J. Kleyna

	in Ymir's blood.			
Saturn XXXIX (Bestla)	Norse primeval goddess, mother of deities, daughter of the giant Bolthorn.	December 13, 2004	Mauna Kea	S. Sheppard, D.C. Jewitt, J. Kleyna
Saturn XL (Farbauti)	Norse storm giant, father of Loki.	December 12, 2004	Mauna Kea	S. Sheppard, D.C. Jewitt, J. Kleyna
Saturn XLI (Fenrir)	Norse monstrous wolf, son of Loki and the giantess Angurboda, father of Hati and Skoll.	December 13, 2004	Mauna Kea	S. Sheppard, D.C. Jewitt, J. Kleyna
Saturn XLII (Fornjot)	Early Norse storm giant, father of Aegir, Kari, and Loge.	December 12, 2004	Mauna Kea	S. Sheppard, D.C. Jewitt, J. Kleyna
Saturn XLIII (Hati)	Gigantic Norse wolf, twin of Skoll.	December 12, 2004	Mauna Kea	S. Sheppard, D.C. Jewitt, J. Kleyna
Saturn XLIV (Hyrrokkin)	Norse giantess who launched Balder's funeral ship. (Spelling changed from Hyrokkin.)	December 12, 2004	Mauna Kea	S. Sheppard, D.C. Jewitt, J. Kleyna
Saturn XLV (Kari)	Norse wind giant.	January 4, 2006	Mauna Kea	S. Sheppard, D.C. Jewitt, J. Kleyna
Saturn XLVI (Loge)	Norse fire giant, son of Fornjot.	January 5, 2006	Mauna Kea	S. Sheppard, D.C. Jewitt, J. Kleyna
Saturn XLVII	Gigantic Norse wolf,	January 5,	Mauna Kea	S. Sheppard, D.C. Jewitt, J.

(Skoll)	twin of Hati.	2006		Kleyna
Saturn XLVIII (Surtur)	Norse leader of the fire giants.	January 5, 2006	Mauna Kea	S. Sheppard, D.C. Jewitt, J. Kleyna
Saturn XLIX (Anthe)	One of the Alkyonides, the seven beautiful daughters of the Giant Alkyoneos.	May 30, 2007		Cassini Imaging Science Team
Saturn L (Jarnsaxa)	Norse giantess and Thor's lover.	January 5, 2006	Mauna Kea	S. Sheppard, D.C. Jewitt, J. Kleyna
Saturn LI (Greip)	Norse giantess.	January 5, 2006	Mauna Kea	S. Sheppard, D.C. Jewittt, J. Kleyna
Saturn LII (Tarqeq)	Inuit moon spirit.	January 16, 2007	Mauna Kea	S. Sheppard, D.C. Jewittt, J. Kleyna
Saturn LIII (Aegaeon)	Greek hundred-armed giant, called Briareus by the gods.	August 15, 2008		Cassini Imaging Science Team

Uranus and its Moons
Satellites in the uranian system are named for characters from Shakespeare's plays and from Pope's "Rape of the Lock."

Body	Description	Date of Discovery	Discovery Location	Discoverer
Uranus	Several astronomers, including Flamsteed and Le Monnier, had observed Uranus earlier but had recorded it as a fixed star. Herschel tried unsuccessfully to name his discovery "Georgian Sidus"	March 13, 1781	Bath	W. Herschel

	after George III; the planet was named by Johann Bode in 1781 after the ancient Greek deity of the sky Uranus, the father of Kronos (Saturn) and grandfather of Zeus (Jupiter).			
Uranus I (Ariel)	Named by John Herschel for a sylph in Pope's "Rape of the Lock."	October 24, 1851	Liverpool	W. Lassell
Uranus II (Umbriel)	Umbriel was named by John Herschel for a malevolent spirit in Pope's "Rape of the Lock."	October 24, 1851	Liverpool	W. Lassell
Uranus III (Titania)	Named by Herschel's son John in early 19th century for the queen of the fairies in Shakespeare's "A Midsummer Night's Dream."	January 11, 1787	Slough	W. Herschel
Uranus IV (Oberon)	Named by Herschel's son John in early 19th century for the king of the fairies in Shakespeare's "A Midsummer Night's Dream."	January 11, 1787	Slough	W. Herschel
Uranus V (Miranda)	Named by Kuiper for the heroine of Shakespeare's "The Tempest."	February 16, 1948	Fort Davis	G.P. Kuiper
Uranus VI (Cordelia)	Daughter of Lear in Shakespeare's "King Lear."	January 20, 1986	Voyager 2	Voyager Science Team
Uranus VII (Ophelia)	Daughter of Polonius, fiance of Hamlet in Shakespeare's "Hamlet, Prince of Denmark."	January 20, 1986	Voyager 2	Voyager Science Team

Exploring and Settling Our Huge Solar System

Uranus VIII (Bianca)	Daughter of Baptista, sister of Kate, in Shakespeare's "Taming of the Shrew."	January 23, 1986	Voyager 2	Voyager Science Team
Uranus IX (Cressida)	Title character in Shakespeare's "Troilus and Cressida."	January 9, 1986	Voyager 2	Voyager Science Team
Uranus X (Desdemona)	Wife of Othello in Shakespeare's "Othello, the Moor of Venice."	January 13, 1986	Voyager 2	Voyager Science Team
Uranus XI (Juliet)	Heroine of Shakespeare's "Romeo and Juliet."	January 3, 1986	Voyager 2	Voyager Science Team
Uranus XII (Portia)	Wife of Brutus in Shakespeare's "Julius Caesar."	January 3, 1986	Voyager 2	Voyager Science Team
Uranus XIII (Rosalind)	Daughter of the banished duke in Shakespeare's "As You Like It."	January 13, 1986	Voyager 2	Voyager Science Team
Uranus XIV (Belinda)	Character in Pope's "Rape of the Lock."	January 13, 1986	Voyager 2	Voyager Science Team
Uranus XV (Puck)	Mischievous spirit in Shakespeare's "A Midsummer Night's Dream."	December 30, 1985	Voyager 2	Voyager Science Team
Uranus XVI (Caliban)	Named for the grotesque, brutish slave in Shakespeare's "The Tempest."	September 6, 1997	Palomar	B. Gladman, P. Nicholson, J.A. Burns and J. Kavelaars
Uranus XVII (Sycorax)	Named for Caliban's mother in Shakespeare's "The Tempest."	September 6, 1997	Palomar	P. Nicholson, B. Gladman, J. Burns and J. Kavelaars

Uranus XVIII (Prospero)	Named for the rightful Duke of Milan in "The Tempest."	July 18, 1999	Mauna Kea	M. Holman, J. Kavelaars, B. Gladman, J.-M. Petit, and H. Scholl
Uranus XIX (Setebos)	Setebos was a new-world (South American) deity's name that Shakespeare popularized as Sycorax's god in "The Tempest."	July 18, 1999	Mauna Kea	J. Kavelaars, B. Gladman, M. Holman, J.-M. Petit, and H. Scholl
Uranus XX (Stephano)	Named for a drunken butler in "The Tempest."	July 18, 1999	Mauna Kea	B. Gladman, M. Holman, J. Kavelaars, J.-M. Petit, and H. Scholl
Uranus XXI (Trinculo)	A jester in Shakespeare's "The Tempest."	August 13, 2001	Cerro Tololo	M. Holman, J.J. Kavelaars and D. Milisavljevic
Uranus XXII (Francisco)	A lord in "The Tempest."	August 13, 2001	Cerro Tololo	J. Kavelaars, M. Holman, D. Milisavljevic, and T. Grav
Uranus XXIII (Margaret)	A gentlewoman attending on Hero from "Much Ado About Nothing."	August 29, 2003	Mauna Kea	S.S. Sheppard, D.C. Jewitt
Uranus XXIV (Ferdinand)	Son of the King of Naples in "The Tempest."	August 13, 2001	Cerro Tololo	D. Milisavljevic, M. Holman, J. Kavelaars, and T. Grav
Uranus XXV (Perdita)	Daughter of Leontes and Hermione in "The Winter's Tale."	January 18, 1986	Voyager 2	E. Karkoschka

| Uranus XXVI (Mab) | The fairies' midwife in "Romeo and Juliet." | August 25, 2003 | Hubble Space Telescope | M.R. Showalter and J.J. Lissauer |
| Uranus XXVII (Cupid) | A character in "Timon of Athens." | August 25, 2003 | Hubble Space Telescope | M.R. Showalter and J.J. Lissauer |

Neptune and its Moons

Satellites in the neptunian system are named for characters from Greek or Roman mythology associated with Neptune or Poseidon or the oceans. Irregular satellites are named for the Nereids, the daughters of Nereus and Doris, and the attendants of Neptune.

Body	Description	Date of Discovery	Discovery Location	Discoverer
Neptune	Neptune was "predicted" by John Couch Adams and Urbain Le Verrier who, independently, were able to account for the irregularities in the motion of Uranus by correctly predicting the orbital elements of a trans- Uranian body. Using the predicted parameters of Le Verrier (Adams never published his predictions), Johann Galle observed the planet in 1846. Galle wanted to name the planet for Le Verrier, but that was not acceptable to the international astronomical community. Instead, this planet is named for the Roman god of the sea.	September 23, 1846	Berlin	J.G. Galle
Neptune I (Triton)	Triton is named for the sea-god son of Poseidon (Neptune) and Amphitrite. The first suggestion of the name Triton has been attributed to the French astronomer Camille Flammarion.	October 10, 1846	Liverpool	W. Lassell

Exploring and Settling Our Huge Solar System

Neptune II (Nereid)	The Nereids were the fifty daughters of the sea god Nereus and Doris and were attendants of Poseidon (Neptune).	May 1, 1949	Fort Davis	G.P. Kuiper
Neptune III (Naiad)	The name of a group of Greek water nymphs who were guardians of lakes, fountains, springs, and rivers.	August 1989	Voyager 2	Voyager Science Team
Neptune IV (Thalassa)	Greek sea goddess. Mother of Aphrodite in some legends; others say she bore the Telchines.	August 1989	Voyager 2	Voyager Science Team
Neptune V (Despina)	Daughter of Poseidon (Neptune) and Demeter.	July 1989	Voyager 2	Voyager Science Team
Neptune VI (Galatea)	One of the Nereids, attendants of Poseidon.	July 1989	Voyager 2	Voyager Science Team
Neptune VII (Larissa)	A lover of Poseidon. After the discovery by Voyager, it was established that an occultation of a star by this satellite had been fortuitously observed in 1981 by H. Reitsema, W. Hubbard, L. Lebofsky, and D. J. Tholen.	July 1989	Voyager 2	Voyager Science Team
Neptune VIII (Proteus)	Greek sea god, son of Oceanus and Tethys.	June 1989	Voyager 2	Voyager Science Team
Neptune IX (Halimede)	One of the Nereids, the fifty daughters of Nereus and Doris.	August 14, 2002	Cerro Tololo	M. Holman, J. Kavelaars, T. Grav, W. Fraser, and D.

Body	Description	Date of Discovery	Discovery Location	Discoverer
				Milisavljevic
Neptune X (Psamathe)	One of the Nereids, lover of Aeacus and mother of Phocus.	August 29, 2003	Mauna Kea	S.S. Sheppard, D.C. Jewitt, and J. Kleyna
Neptune XI (Sao)	One of the Nereids, the fifty daughters of Nereus and Doris.	August 14, 2002	Cerro Tololo	T. Grav, M. Holman, J. Kavelaars, W. Fraser, and D. Milisavljevic
Neptune XII (Laomedeia)	One of the Nereids, the fifty daughters of Nereus and Doris.	August 13, 2002	Cerro Tololo	J. Kavelaars, M. Holman, T. Grav, W. Fraser, and D. Milisavljevic
Neptune XIII (Neso)	One of the Nereids, the fifty daughters of Nereus and Doris.	August 14, 2002	Cerro Tololo	M. Holman, J. Kavelaars, T. Grav, W. Fraser, and D. Milisavljevic
Neptune XIV (Hippocamp)	Mythical seahorse in Greek mythology, a symbol of Poseidon.	July 15, 2013	Hubble Space Telescope	M. Showalter, I. de Pater, T. Grav, J. J. Lissauer, and R. S. French

Pluto and its Moons

Satellites in the plutonian system are named for characters and creatures in the myths surrounding Pluto (Greek Hades) and the classical Greek and Roman Underworld.

Body	Description	Date of Discovery	Discovery Location	Discoverer

Exploring and Settling Our Huge Solar System

(134340) Pluto	Pluto was discovered at Lowell Observatory in Flagstaff, AZ during a systematic search for a trans-Neptune planet predicted by Percival Lowell and William H. Pickering. Named after the Roman god of the underworld who was able to render himself invisible.	January 23, 1930	Flagstaff	C.W. Tombaugh
(134340) Pluto I (Charon)	Named after the Greek mythological boatman who ferried souls across the river Styx to Pluto for judgement.	April 13, 1978	Flagstaff	J.W. Christy
(134340) Pluto II (Nix)	Goddess of darkness and night, mother of Charon. (Nix is the Egyptian spelling of the Greek name Nyx.)	May 15, 2005	Hubble Space Telescope	H.A. Weaver, S.A. Stern, M.J. Mutchler, A.J. Steffl, M.W. Buie, W.J. Merline, J.R. Spencer, E.F. Young, and L.A. Young
(134340) Pluto III (Hydra)	In Greek mythology, terrifying monster with the body of a serpent and nine heads that guarded the underworld.	May 15, 2005	Hubble Space Telescope	H.A. Weaver, S.A. Stern, M.J. Mutchler, A.J. Steffl, M.W. Buie, W.J. Merline, J.R. Spencer, E.F. Young, and L.A. Young
(134340) Pluto IV (Kerberos)	In Greek mythology, the many-headed dog that guarded the entrance to the underworld.	June 28, 2011	Hubble Space Telescope	M.R. Showalter, D.P. Hamilton, S.A. Stern, H.A. Weaver, A.J. Steffl, and L.A. Young
(134340) Pluto V	Greek goddess who ruled over the underworld river	June 26, 201		

(Styx)	also named Styx.